Book 8

Human Sexuality

REPRODUCTION IN MAMMALS

Book 8

Human Sexuality

EDITED BY

C. R. AUSTIN
Fellow of Fitzwilliam College,
Charles Darwin Professor of Animal Embryology,
University of Cambridge

AND

R. V. SHORT, FRS
Director of the Medical Research Council
Unit of Reproductive Biology,
Honorary Professor,
University of Edinburgh

ILLUSTRATIONS BY JOHN R. FULLER

CAMBRIDGE UNIVERSITY PRESS
CAMBRIDGE
LONDON · NEW YORK · NEW ROCHELLE
MELBOURNE · SYDNEY

Published by the Press Syndicate of the University of Cambridge
The Pitt Building, Trumpington Street, Cambridge CB2 1RP
32 East 57th Street, New York, NY 10022, USA
296 Beaconsfield Parade, Middle Park, Melbourne 3206, Australia

First published 1980

Printed in Great Britain at the
University Press, Cambridge

British Library Cataloguing in Publication Data
Reproduction in mammals.
Book 8: Human sexuality
1. Mammals – Reproduction
I. Austin, Colin Russell
II. Short, Roger Valentine
599′.01′6   QL739.2   80-40038
ISBN 0-521-22361-X
ISBN 0-521-29461 Pbk

Contents

Contents

Contributors to Book 8

C. R. Austin,
Physiological Laboratory,
Downing Street,
Cambridge, CB2 3EG

J. Bancroft,
M.R.C. Reproductive Biology Unit,
Centre for Reproductive Biology,
37 Chalmers Street,
Edinburgh, EH3 9EW

G. R. Dunstan,
King's College London,
Strand,
London, WC2R 2LS

R. Green,
Department of Psychiatry and Behavioral Science,
State University of New York,
Stony Brook,
New York 11794,
USA

M. Schofield,
28 Lyndhurst Gardens,
Hampstead,
London, NW3 5NW

R. V. Short,
M.R.C. Reproductive Biology Unit,
Centre for Reproductive Biology,
37 Chalmers Street,
Edinburgh, EH3 9EW

Books in this series

Book 1. Germ Cells and Fertilization
Primordial germ cells.   T. G. Baker
Oogenesis and ovulation.   T. G. Baker
Spermatogenesis and the spermatozoa.   V. Monesi
Cycles and seasons.   R. M. F. S. Sadleir
Fertilization.   C. R. Austin

Book 2. Embryonic and Fetal Development
The embryo.   Anne McLaren
Sex determination and differentiation.   R. V. Short
The fetus and birth.   G. C. Liggins
Manipulation of development.   R. L. Gardner
Pregnancy losses and birth defects.   C. R. Austin

Book 3. Hormones in Reproduction
Reproductive hormones.   D. T. Baird
The hypothalamus.   B. A. Cross
Role of hormones in sex cycles.   R. V. Short
Role of hormones in pregnancy.   R. B. Heap
Lactation and its hormonal control.   Alfred T. Cowie

Book 4. Reproductive Patterns
Species differences.   R. V. Short
Behavioural patterns.   J. Herbert
Environmental effects.   R. M. F. S. Sadleir
Immunological influences.   R. G. Edwards
Aging and reproduction.   C. E. Adams

Book 5. Artificial Control of Reproduction
Increasing reproductive potential in farm animals.   C. Polge
Limiting human reproductive potential.   D. M. Potts
Chemical methods of male contraception.   Harold Jackson
Control of human development.   R. G. Edwards
Reproduction and human society.   R. V. Short
The ethics of manipulating reproduction in man.   C. R. Austin

Book 6. The Evolution of Reproduction
The development of sexual reproduction.   S. Ohno
Evolution of viviparity in mammals.   G. B. Sharman
Selection for reproductive success.   P. A. Jewell
The origin of species.   R. V. Short
Specialization of gametes.   C. R. Austin

Preface

This volume of *Reproduction in Mammals* is intended to meet the needs of undergraduates reading zoology, biology, animal behaviour, physiology, psychology, medicine and psychiatry, and as a source of information for advanced students and research workers. At the same time, specialist's jargon has been kept to a minimum and the text should prove fully comprehensible to any serious enquirer. The contents list of the books in this series are set out on the previous two pages.

We have attempted to look at human sexuality from a number of very different viewpoints, including those of the biologist, comparative anatomist, psychologist, psychiatrist and moral philosopher, in the belief that this breadth of coverage is necessary if one is to deal with such a sensitive and emotive subject in a balanced, informative and constructive manner.

1 The origins of human sexuality
R. V. Short

There are two problems to be overcome if we are to attempt to take an objective look at normal human sexual behaviour. The first is that we are all so intimately involved in sex ourselves that our own personal experiences are bound to cloud our judgements. The second is that human sexual behaviour is infinitely variable, within the cultural constraints imposed upon it. Thus we can find examples of polyandry, polygyny, monogamy, serial monogamy, promiscuity, homosexuality or even celibacy being regarded as the norm for a particular community. Faced with such a diversity of sexual practices, how can we ever reach any objective conclusions about the innate norms for our species? Clearly we cannot start from the presumption that the Western, Christian ethic of lifelong monogamy is the only natural, decent thing to do.

What we need is some point of reference that is independent of our own cultural upbringing. In this chapter, I would like to develop the theme that we may learn something about ourselves by comparison with our closest living relatives, the gorilla, chimpanzee and orang-utan. Although separated from them by several million years of independent evolution, and adapted to suit widely different environments, we still share a very similar genotype. As pointed out in Book 6, Chapter 4, there are striking similarities between the gross structure and banding patterns of our chromosomes. And the genetic distance between man and the three Great Apes, as judged by amino acid sequencing of blood proteins, is incredibly small, comparable to that between closely related species of the same genus in other groups of organisms, e.g. *Drosophila*. But if our structural genes are so similar, we know that there has been great diversity in the selection of those regulatory genes that control the phenotypic expression of characters. One of the fundamental tenets of biology is that form reflects function, and if we can compare and contrast the reproductive anatomies of man and the Great Apes, this may lead us to some valuable insights into the selective forces that have shaped our sexuality during the later stages of our evolution.

Some people seem to find such an approach quite unacceptable, since it smacks of the anthropomorphic. But nobody seems to get upset when archaeologists attempt to piece together our evolutionary history from careful reconstructions of dated fossil remains. Nobody disputes the fact that the dentition of our prehistoric ancestors provides vital clues about our changing dietary habits. The fact that man has the greatest cranial capacity of all the primates, both in relative and absolute terms, is generally agreed to be one of the outstanding features of human evolution. Therefore, why should we not speculate about the significance of the differences that exist in the reproductive anatomy of man and the Great Apes, and, like any good tipster, attempt to guess at performance by studying form?

THE SIZE OF THE GONADS

Let us begin by considering the ovary and the testis, and the basic question of what it is that primarily determines gonadal size. Since the gonads are endocrine glands releasing steroid hormones into the systemic circulation, the volume of endocrine tissues, be they Graafian follicles, corpora lutea or Leydig cells, is likely to vary in relation to total body mass.

But the principal function of the gonads is to release gametes to the exterior, and here there are fundamental differences in the constraints upon an ovary and a testis. As we have seen (Book 1, Chapter 2), only a minute proportion of the total population of oocytes in the ovary at birth are ever ovulated; the majority perish through atresia. Since the oocytes of all mammals are remarkably similar in size (Book 6, Chapter 5), major differences between species in body size, life span, or ovulation rate do not therefore necessitate comparable differences in the volume of germinal tissue in the ovary. Even after ovulation, the success of the fimbria in picking up the shed oocyte is presumably unaffected by the size of the animal.

Taking all these facts into consideration, we might expect the volume of endocrine tissue in the ovary to vary with body size, but the volume of germinal tissue to remain relatively constant, so that ovary:body weight ratios would be similar between species. This certainly seems to be true for man and the Great Apes (see Table 1-1).

TABLE 1-1. The weight of the gonads as a percentage of total body weight in adult non-pregnant apes and man. (From Short, 1979)

Species	Testes	Ovaries
Gorilla	0.017	0.012
Orang-utan	0.048	0.006
Chimpanzee	0.269	0.010
Man	0.079	0.014

But the factors governing testicular size are entirely different, and the relative size of the testis may provide some particularly useful clues about the sexual behaviour of the species concerned. Although the size of the individual spermatozoon is relatively constant between mammals (Book 6, Chapter 5), there are enormous differences between species both in the number of spermatozoa in an ejaculate, and in the ejaculatory frequency. Large animals have to produce more spermatozoa because of the greater volume of the female tract through which the ejaculate becomes distributed. But even more important than body size is the frequency of ejaculation. In a polygynous species with a restricted rutting period, the male may have to adapt to a very high copulatory frequency for a limited period of time. In promiscuous species, the position is complicated still further by

3

gamete selection; if several different males mate with a female at one oestrus, the male who deposits the largest number of spermatozoa, other things being equal, will be the most likely to sire the offspring, and hence will be favoured by selection. At the other extreme in monogamous species where the female only allows copulation when she is in oestrus, the copulatory frequency may be extremely low. How can the testis respond to these varying demands that may be made upon it?

Rupert Amman has shown that the rate of sperm production per gramme of testicular tissue is relatively constant across a range of species, from the rhesus monkey to the rat, so the main way of increasing the daily rate of sperm production is by increasing the volume of seminiferous tubular tissue. And since seminiferous tubule diameters are similar in all mammals, this volume increase can only be achieved by increasing the number and/or length of the tubules. Peter Hayward and John Shire have shown that in the mouse there is a Y-linked gene that determines the length of the seminiferous tubules. Since Susu Ohno (Book 6, Chapter 1) has taught us that there has been extreme evolutionary conservation of sex-linked genes across the species, it is likely that this same Y-linked gene for testis size would be present in all mammals. If a gene is Y-linked it is only present in the male; since it is always in the haploid state, it will respond very rapidly to selection pressure. Therefore, it seems likely that the mammalian testis could easily adapt to demands for increased sperm production. Although the mass of germinal and Leydig tissue will vary in relation to total body mass, there will be an additional major variation in germinal tissue which will reflect variations in mating behaviour and copulatory frequency.

If we take a look at the testes:body weight ratios in man and the Great Apes, enormous species differences do indeed exist, with the chimpanzee having by far the largest testes in both relative and absolute terms, and the gorilla by far the smallest (see Table 1-1 and Fig. 1-1). Indeed, the gorilla has the lowest testes:body weight ratio of all the primates. Why should this

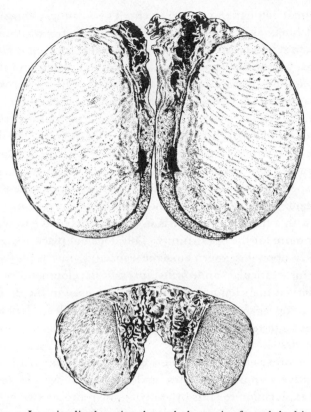

Fig. 1-1. Longitudinal section through the testis of an adult chimpanzee (*above*) and an adult mountain gorilla (*below*), both × ½. Since a male chimpanzee weighs only about 50 kg, and a male gorilla about 200 kg, the enormous differences in both relative and absolute testicular size between these two species can readily be appreciated. This testicular hypertrophy in the chimpanzee is probably related to its high copulatory frequency. (From G. B. Wislocki, *Journal of Mammalogy* **23**, 281 (1942).)

monstrous beast be so poorly endowed, and why should our own testicular size pale into insignificance when compared to that of the diminutive chimpanzee?

Studies of the reproductive behaviour of wild mountain gorillas in Africa carried out initially by George Schaller, and more recently by Sandy Harcourt and Kelly Stewart, have shown that copulation is a rare event. This is simply explained if one thinks about the make-up of a gorilla troop, with its dominant silverbacked male who is responsible for almost all the matings, one or two subordinate blackbacked males, and three to six sexually mature females. Copulation only occurs when there is a female in oestrus, and she is usually the one to initiate the act. But the females rarely come into oestrus; this is because most of a female's reproductive life is occupied by long periods of lactational anoestrus, which effectively keep the births spaced about four years apart. Thus each female in the troop may only experience two or three two-day periods of oestrus every four years. Although the male may copulate every two or three hours with a female when she is in heat, he may have to wait a year or more for the opportunity. This does not place any great demands upon his spermatogenic capacity, but he certainly needs his testicular androgens for the development of his secondary sexual characteristics, which are used in the defence of his troop against the incursions of rival males. Hence his diminutive testes show well developed Leydig cells and sparse tubular tissue.

At the other end of the scale we have the chimpanzee, with his relatively enormous testes, and once again we can find a logical explanation for this by studying the reproductive behaviour of chimpanzees in the wild. Caroline Tutin has carried out just such an investigation in the Gombe Stream Reserve in Tanzania, where the animals live in large, multimale, multifemale social groups with an equal sex distribution and a total population of maybe 30 adults. Although there is competition between males in a variety of social situations, and group defence by the males of a communal territory in which the females reside, the mating system is essentially promiscuous rather than polygynous, since all the males in a troop may take turns in copulating with an oestrous female – a 'gang bang', in common parlance.

Although the birth interval is even longer than in the gorilla, approaching six years, the availability of oestrous females and hence the copulatory frequency is enormously increased in the chimpanzee. There are several reasons for this. In the first place, the duration of oestrus is prolonged to about ten days, compared with two days in the gorilla. Then the promiscuous mating system effectively makes all the females in the troop available to every adult male, thereby increasing harem size to 15 or more. And finally, the pubertal females have one or two years of monthly oestrous cycles before they eventually conceive. It can be calculated that an individual female will spend about 500 days of her life in oestrus, half of them during this pubertal period, although she will only have about seven pregnancies. Thus the majority of acts of copulation are not related to conception, and it would seem that sexual activity has become significant for the social cohesion of the group, as well as for mere procreation.

We can appreciate that, in the wild, hardly a week will go by without some female chimpanzee in the troop being in heat. Copulation is a brief affair, lasting a mere eight seconds, and each male may copulate many times in one day. This, plus the added fact that gamete selection will favour the male who deposits the most spermatozoa, has presumably led to selection for increased spermatogenic activity and hence the development of very large testes composed mostly of seminiferous tubules.

Table 1-1 shows that human testicular size most closely approximates to that of the orang-utan, so should we look to the orang for some clues about our innate sexuality? Almost certainly not; the orang is a desocialized ape, the male and the female spending the greater part of their reproductive lives apart. This solitary lifestyle has apparently been forced upon the animal by its feeding habits. It is a fruit eater, leading an arboreal existence in the great forests of Borneo and Sumatra, where the density of fruit-bearing trees is so low that a large area is required to meet the nutritional demands of so big an animal. A troop of orangs trying to share the food supply between them would soon starve to death. Biruté Galdikas has studied the sexual behaviour of wild orang-utans in Borneo, and has shown that the female only seeks out the male when she is in oestrus, locating him by his loud vocalizations. However, each female only comes into

7

oestrus every five or six years, because of the long duration of lactational anoestrus. Thus although each male has access to more than one female, the male's relatively small testes certainly confirm that copulation is not nearly as frequent as in the chimpanzee, but perhaps more frequent than in the gorilla.

And so we come to man himself. The size of the human testis is nothing to boast about: the average weight in 140 Caucasians was found to be $21.6 \pm$ S.E.0.4 g for the right side and 20.4 ± 0.5 g for the left, a significant asymmetry which is not related to whether you are right- or left-handed! Curiously, there appear to be marked racial differences in testis size; the mean weight of the testes in a population of 100 Chinese men from Hong Kong was 10.0 ± 0.3 g for the right testis, and 9.4 ± 0.3 g for the left. Some of this difference may be merely a reflection of racial differences in overall body size between Caucasians and Orientals. Certainly, we know that in bulls there is a significant correlation of 0.58 between scrotal circumference and body weight. But it would be hard to imagine that differences in body weight could account for all of the observed differences in human testis size, since Oriental men are not on average half the size of Caucasians.

It would be interesting to know whether these racial differences in human testis size are in any way related to differences in copulatory frequencies, and whether they are reflected in differences in sperm production rates; unfortunately, this type of information is entirely lacking. However, there is some indirect evidence to suggest that this Oriental–Caucasian difference may be autosomal rather than sex-linked in its origins, and that it may be reflected in differences in the female population as well. The argument runs like this. In addition to the Y-linked gene for testis size in the mouse, there must also be autosomal determinants, since Roger Land in Edinburgh showed that if you select male mice for large and small testis size, the females of these lines show high and low ovulation rates, respectively. Presumably one has selected for some autosomal gene or genes that influence a reproductive characteristic common to both sexes, such as

8

pituitary gonadotrophin secretion, or gonadal sensitivity to gonadotrophic stimulation. So, do Oriental women show any diminution in ovulation rate that might be related to the reduced testicular size in Oriental men? Indeed they do. Although differences in ovulation rate are difficult to determine in an essentially monotocous species like man, one useful index is the incidence of dizygotic (two-egg) twinning in the population: Oriental women have the lowest dizygotic twinning rate of any ethnic group.

TABLE 1-2. The weight of the testes and the sperm content of the ejaculate in adult apes and men

Species	Weight of testes (g)	Sperm content of ejaculate ($\times 10^{-6}$)
Gorilla	28	51
Orang-utan	35	67
Chimpanzee	119	603
Man	42	253

Returning once more to the argument about the relationship between testicular size and sperm production rate, it can be appreciated that the latter is an extremely difficult parameter to measure experimentally. The total number of spermatozoa produced per ejaculate is a poor index of the total spermatogenic reserve in the epididymis, although it is true that the sperm content of the ejaculate does seem to be roughly proportional to testicular size (see Table 1-2). Rupert Amman has calculated that in the rhesus monkey the daily rate of sperm production is 23×10^6 spermatozoa per gramme of testicular tissue. If we use the same figure for the chimpanzee, this gives a daily production of 2737×10^6 spermatozoa, sufficient for at least four ejaculates a day. The corresponding figure for the human testis is 4.4×10^6 spermatozoa per gramme of testis, giving a daily production of

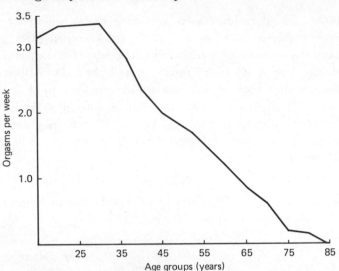

Fig. 1-2. Frequency of orgasm in American men by age. (From A. C. Kinsey, W. B. Pomeroy and C. E. Martin. *Sexual Behavior in the Human Male*. Philadelphia; Saunders (1948).)

176×10^6 spermatozoa, which is *less* than the number in one ejaculate (see Table 1-2).

We can also get an indirect indication of the sperm production rate in men from the work of Matthew Freund in New York, who showed that men can maintain a normal output of about 250×10^6 spermatozoa per ejaculate at ejaculatory frequencies of 3.5 times per week; it will fall to 60×10^6 per ejaculate at an ejaculatory frequency of 8.6 times per week, and 31×10^6 at a frequency of 16.9 times per week. Of course, we do not know how low the sperm output has to fall before fertility becomes seriously compromised. If we look at known copulatory frequencies by age for North American men (Fig. 1-2), our normal sexual performance would seem to be well within our spermatogenic capacity but our testes would be ill-adapted to cope with the multiple daily 'gang bangs' of the chimpanzee.

We are apparently the only mammal in which the female has forsaken the periodic behavioural phenomenon of oestrus, in which she is inherently attractive and receptive to the male, and exchanged it for a situation in which she is potentially attractive and receptive at all times from adolescence to old age (see Book 6, Chapter 4). Thus almost all human acts of intercourse are entirely unrelated to the possibility of conception. No wonder that several primitive communities such as the Trobriand Islanders and the Australian Aboriginals failed to associate intercourse with conception; it was probably not until man began to domesticate animals, which came into oestrus, copulated, and gave birth to young at a predictable time thereafter, that the 'facts of life' first dawned on him.

THE SIZE OF THE PENIS

Although modesty prevented Charles Darwin from ever referring to the male copulatory organ, it is a vital part of the male's reproductive anatomy, and has its own tale to tell about the sexual behaviour of the species. Take for example the gorilla. The flaccid penis of the male gorilla is invisible, being concealed within the prepuce, which is merely a small orifice on the ventral body wall, hidden by hair. When the prepuce is reflected, the penis is still inconspicuous, being minute in size and of the same black colour as the surrounding skin. Even when fully erect, it

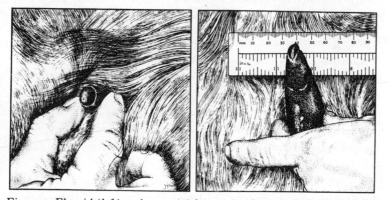

Fig. 1-3. Flaccid (*left*) and erect (*right*) penis of an adult, fertile gorilla, showing the inconspicuous nature of the organ, even when fully erect.

Fig. 1-4. The normal dorso-ventral copulatory position in the gorilla, probably dictated by the small size of the penis. Note the marked sexual dimorphism in body size, and especially in head size.

is only 3 cm long, and barely visible at a distance (Fig. 1-3). This certainly suggests that the gorilla's penis is unlikely to be used in sexual display: indeed, we know that the oestrous female rather than the male initiates the rare copulatory acts. The small penis may also dictate the copulatory position; in the wild, this is invariably dorso-ventral (Fig. 1-4), and it is difficult to see how the male could achieve intromission in a ventro-ventral 'missionary' position.

The situation in the chimpanzee is entirely different. Although the flaccid penis is normally hidden from view by the prepuce, whose orifice is flush with the body wall, this whole area is more

Fig. 1-5. Sexual display of male pygmy chimpanzee, with erect penis, towards an oestrous female with maximal perineal swelling.

conspicuous because of its lack of long hair. The erect penis is about 8 cm in length, and its bright pink colour stands out clearly against the skin (Fig. 1-5). This large organ has clearly been developed for copulatory display; males almost always take the initiative in courtship, and either approach the female with an erect penis and copulate immediately, or else indulge in a much more elaborate display. Copulation normally occurs in the conventional mammalian dorso-ventral position; the tumescent perineum of the oestrous female has presumably necessitated the development of a long, thin, filiform penis for adequate penetration and intromission.

The penis of the orang-utan is completely concealed from view when within the prepuce, whose opening is flush with the body wall, and hidden by the long, coarse red hair. When erect, its pink colour makes it conspicuous, although it is only 4 cm in length, which is very small for so large an animal. For reasons already stated, we know almost nothing of the copulatory displays of orangs in the wild, although copulation has occasionally been observed. The animals copulate in the trees, whilst hanging from branches, so great agility is required, and several copulatory positions are adopted; it is reported that intromission is only achieved with difficulty. In captivity, where the animals are made to copulate on the ground, the mean duration is about 15 minutes.

The most remarkable feature of our own reproductive anatomy is the conspicuousness of the penis, whether flaccid or erect, and of the testes, which are displayed to view in a pendulous scrotum. Even the pubic hair, which in the gorilla and orang acted as additional camouflage for the external genitalia, seems designed in man to draw attention to the genitalia, rather than to conceal them (Fig. 1-6). It is interesting that the first noticeable change at puberty in a boy is enlargement of the testes, and we have been able to show that this is coincident with the first appearance of spermatozoa, determined by examining centrifuged deposits from urine. Thus boys go through a phase when they could be described as 'fertile eunuchs', producing spermatozoa but having external genitalia that are not so very different from those of an adult male gorilla. The fact that pubic hair development and phallic enlargement occurs at puberty emphasizes that these are secondary sexual characteristics that have presumably developed under the influence of Sexual Selection. We must now ask ourselves what selective forces have been at work.

The erect human penis reaches truly enormous proportions by gorilla standards, with a mean length of 13 cm (Fig. 1-7); this is an androgen-dependent development, unrelated to frequency of intercourse, or even to general body build. Perhaps the large size of the erect penis is related to the act of intercourse. As Leonardo da Vinci showed in his magnificent drawing (Fig. 1-8), the size of the penis makes ventro-ventral copulation possible; indeed, this is the normal position for most couples. But the ability to adopt a wide variety of other copulatory positions has presumably enhanced the enjoyment of intercourse for both partners, and may have reinforced the pair-bonding role of human sexuality.

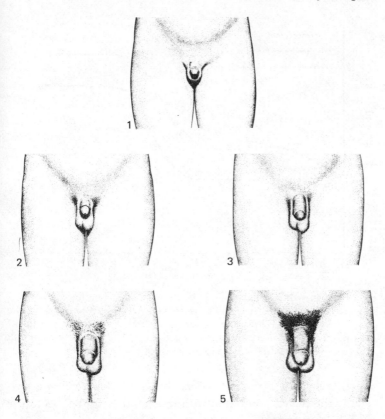

Fig. 1-6. Stages of genital development in a circumcised boy going through puberty. Testicular enlargement is noticeable at stage 2, and this is when spermatozoa are first produced. Pubic hair first appears at stage 3, and the genitalia become fully adult at stage 5. (From W. A. Marshall and J. M. Tanner, *Archives of Diseases of Childhood* **45**, 13 (1970).)

The human penis may also have evolved as an organ of display, and perhaps this has been its principal function in the past. The erect penis has always been regarded as highly erotic and is depicted in rock paintings, carvings, sculptures and drawings from every corner of the globe. But erotic for whom? When Darwin first put forward his ideas about Sexual Selection in 1871, he realized that there were basically two types. Intra-

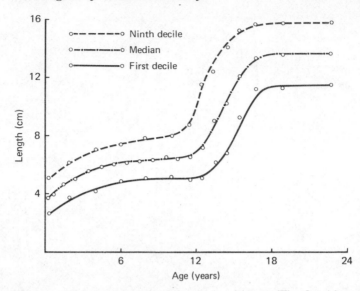

Fig. 1-7. Growth in length of the penis with age. The flaccid penis is measured when fully stretched, as this closely approximates to the length when erect. Whilst spermatozoa are first produced at the age of 12–13, penis growth occurs somewhat later. (From W. A. Schonfeld. *American Journal of Diseases of Childhood* **55**, 535 (1943).)

sexual Selection occurred when the individuals of one sex, usually the males, competed with one another for access to the opposite sex, and this led to the development of large body size, and of weapons of offence, like horns, antlers, large canine teeth, etc. Intersexual Selection, on the other hand, occurred when one sex tried to entice the opposite sex by developing specific adornments, like the tail of the male peacock, or the women's breasts. So how did the human penis develop? Was it for threat and display between males, as has been suggested by Wolfgang Wickler, or was it to make the man more attractive to his spouse, and to enhance the enjoyment of the copulatory act? The latter seems the more plausible explanation since the one outstanding fact about normal human heterosexual behaviour in all known communities is that it usually takes place between consenting couples in private. There can be few occasions when males would

Fig. 1-8. Leonardo da Vinci's drawing of human intercourse, made in about 1492–4, and entitled 'I display to men the origin of the second – first or perhaps second – cause of their existence'. Like those before him, Leonardo believed that the thicker part of the semen, carrying the animal spirit or soul of the future embryo, was derived from the spinal cord. The thinner part was derived from the testicles and the bladder. The vein leaving the top of the uterus was thought to carry the blood of the retained menses up to the breasts for the formation of milk – a neat explanation for lactational amenorrhoea. (Reproduced by gracious permission of Her Majesty the Queen).

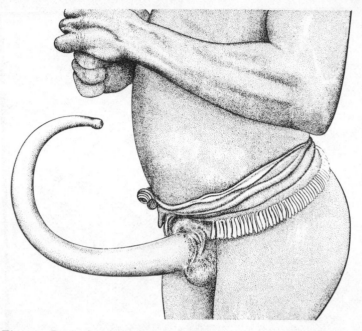

Fig. 1-9. Penis sheath made out of a gourd, being worn by a Telefomin native from New Guinea. Rather than having any phallic significance, penis sheaths were worn out of modesty and decorum by those who had been circumcised, in order to conceal the erotic glans penis from view. (From P. J. Ucko, *Proceedings of the Royal Anthropological Institute*, 1969, p. 27.)

normally be in the habit of displaying the fully erect penis to other males. However, the penis is such a powerful symbol of sexuality that we find a variety of fascinating cultural practices associated with it.

The simplest of these is circumcision. The highly vascularized glans penis is the most erotic part of the organ, normally concealed from view by the prepuce. Many primitive tribes all over the world practised circumcision ceremonies at the time of penile enlargement during puberty; the revealed glans was thus a further reminder to all that the boy had now entered manhood. However, accentuating the erotic role of the penis in this way may have led to embarrassment, and a need to conceal the glans

18

from view once more. This was achieved in many areas of the world by the use of a penis sheath, constructed of bamboo, gourd, horn, shell, basket work, or even metal (Fig. 1-9). Peter Ucko has made a detailed study of this practice, and has shown that the early anthropologists who encountered such tribes for the first time were mistaken in regarding these sheaths as highly erotic phallic symbols. Although they were often elaborately and beautifully decorated, there seems little doubt that in most societies they had remarkably few phallic connotations, and instead were symbols of modesty and decorum. Indeed, the sheaths were often attached to the wearer by threading the flaccid penis, like a bootlace, through a series of apertures at the base of the sheath, thus rendering penile erection impossible, and any intimation of an erection downright painful.

The origins of male circumcision are unknown; it is certainly a very ancient practice, and some of the earliest Egyptian mummies show evidence of the operation. It is interesting to see how in recent years attempts have been made to justify the operation in Western countries on medical grounds. Removal of the prepuce was thought to improve penile hygiene, thereby reducing the risk of penile cancer in the man, or of carcinoma of the cervix in his wife, and also to reduce the risk of venereal infection. However, there seems to be little or no factual support for any of these ideas. In his classical article entitled 'The fate of the foreskin', Douglas Gairdner, a Cambridge paediatrician, pointed out some of the problems of the operation. In man, as in animals, the prepuce is partially adherent to the glans at birth. These adhesions break down gradually during the first few years of life, and in animals we know that this separation can be arrested by castration, so presumably it is androgen-dependent. To practise ritual circumcision, as some Jewish sects do, within a few weeks of birth before preputial separation has occurred is to invite trouble. And although it is a minor operation, it is not without risk of severe haemorrhage, or even death. In the 1940s about 16 children a year died of the operation in England and Wales alone, with most of the deaths occurring in children who were circumcised under one year of age. Surely that is a heavy price to pay for what is, after all, only a primitive ritual mutilation ceremony?

Other forms of penile mutilation have been practised besides circumcision. Peculiar to the Australian Aboriginals is subincision. This was usually performed later on in puberty, some time after circumcision, and entailed making a slit in the urethra from the urethral orifice for a variable distance back towards the scrotum, thus creating an artificial hypospadias. Since the corpus cavernosum urethrae which surrounds the urethra is highly vascular, one shudders to think of the haemorrhage that would have ensued in an adult man; no wonder the Aboriginals called it 'mika', meaning 'the terrible rite'. Some Westerners have fancifully imagined that this was used as a novel form of contraception, whereby the semen would escape from the root of the penis before entering the vagina. However, there is apparently no truth in this suggestion; the fertility of these subincised men seems to have been unimpaired, and semen did not normally escape at intercourse. Why the operation should have been unique to the Aboriginals is unknown, although it has been suggested that it may be related to kangaroo worship; certainly the male kangaroo does have a slight degree of hypospadias.

With the adoption of clothing, some aspects of penile display were sometimes transferred to the mode of dress. In Europe in the late fourteenth century, the codpiece came into use. Although initially it was probably designed to preserve decorum at a time when the jerkin and doublet were becoming shorter, and the hose tighter, by the fifteenth and sixteenth centuries it had frequently become an object of obtrusive display (Fig. 1-10). In design it often simulated the male genitalia, and attention was drawn to it by ribbons and surrounding embroidery. Its popularity waned in the seventeenth century as tight hose gave way to buttoned breeches.

We can gain an interesting insight into present-day attitudes to the penis by considering the dreaded disease of Koro – much

EN·EXPRESSA·VIDES·HENRICI·REGIS·IMAGO
QVA·FVIT·OCTAVI·MVSIS·HOC·STRVXT·ASYLVM
MAGNIFICE·CVMTER·DENOS·REGNASSET·ET·OCTO
ANNOS·QVIS·MAJOR·REGEM·LABOR·VLTIMVS·ORNET

EX·DONO·ROBERTI·BEAVMONT·SACRE·THEOLOGIE·PROFESSORIS·ET·HVIVS·COLLEGII·MAGISTRI·Aº1567

Fig. 1-10. Henry VIII in court garb which includes a distinctive codpiece. (Drawn from a portrait by Hans Eworth after Holbein which hangs in the Dining Hall of Trinity College, Cambridge. By kind permission of the Master and Fellows.)

feared in the Far East. Men become literally terrified by the thought that their penis will retract into the abdomen, gradually disappearing from view. Maybe it is merely the consequence of obesity, but one wonders whether this fear may not have given rise to the widespread use of aphrodisiac preparations in Oriental communities. There is no evidence that any of the orally administered aphrodisiacs, such as deer antler in velvet, has any action whatsoever, other than a placebo effect. However, there may be a more rational basis for the use of rhino horn, the demand for which is so great that it will almost certainly lead to the near extinction of all species of rhinoceros within a very few years. Rhino horn is in fact a keratinous structure, analogous to a clumped mass of hair. When ground down to a powder and applied topically to the penis, it acts as an intense irritant, so that the wearer can think of little else. It is a pity that we could not persuade people to adopt some substitute, like shredded horse hair, thereby saving the rhinoceros from needless slaughter.

The law of the land also has its tale to tell with respect to present-day attitudes to the penis. In these days of sexual equality, one might have expected the abolition of legislation that promoted sexual discrimination. Whilst this has been true of many recent legal reforms, it is interesting to note that in English law, indecent exposure remains an exclusively male offence. It is defined as display of the penis, whether flaccid or erect, towards somebody of the opposite sex, and is a punishable offence. It cannot be committed by a female, since in law she has nothing indecent to expose. It is the commonest of all sexual offences, and there are about 3000 convictions a year in England and Wales alone, mostly in the 20–40-year age group.

Officialdom can occasionally show a sense of humour in matters sexual. Carved into the chalk on a hillside above the village of Cerne Abbas in Dorset is the vast figure of the Cerne Giant, which dates from about the second century A.D. and probably represents the Romano–British deity Hercules-Ognios (see Fig. 1-11). On 14 November 1932 an outraged citizen of Dorset, one Walter L. Long, wrote to the Home Office:

22

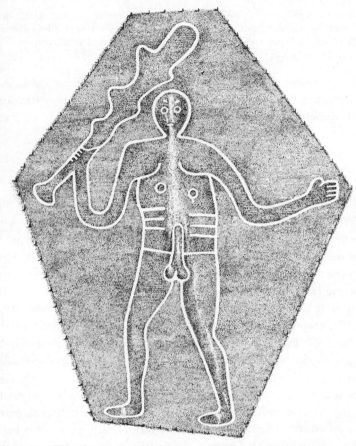

Fig. 1-11. The Cerne Giant, a Romano–British figure carved into the chalk hillside above the village of Cerne Abbas, in Dorset.

Were the Cerne Giant converted into a simple nude, no exception would be taken to it. It is its impassioned obscenity that offends all who have the interest of the rising [*sic*] generation at heart, and I, we, appeal to you to make this figure conform to our Christian standards of civilization.

There is on file at the Public Record Office a Home Office minute, written by an L. Cecil Yates:

This is a serious charge of indecency against a scheduled prehistoric national monument, made, with apparent deliberation, after a lapse of

2000 or 3000 years...What does the complainant want us to do? Commit a nameless outrage? We cannot contemplate that. Plant a small grove of fig trees (on measurements, hardly less would suffice) in a strategic position?

In the event, the Home Office eventually replied to Mr Long, informing him that since the Giant of Cerne was a national monument, vested in the National Trust, the Secretary of State could not see his way to take any action in the matter.

DEVELOPMENT OF THE FEMALE GENITALIA

The striking differences we have seen between man and the Great Apes in the form of the male external genitalia are also reflected in the female genitalia. The most notable example is in the chimpanzee, where the female advertises her sexual state to the male by means of a pronounced swelling and hyperaemia of the labial and circumanal region (Figs. 1-5 and 1-12). The swelling is maximally developed in the middle of the menstrual cycle, around the time of ovulation, and subsides rapidly thereafter; the gross volume change may amount to as much as 1400 ml during the course of the cycle. Females will normally only permit copulation during tumescence, when they are also most attractive to males. Since the mating system is essentially a promiscuous one, the female has nothing to lose by this flagrant visual advertisement of her reproductive state, and as we have already mentioned, group sex may be a factor in promoting group cohesion in this species. The female gorilla is the only other ape to show any sign of sexual swelling during the menstrual cycle, although it is not at all pronounced. No such changes are present in the female orang, and it is difficult to see how visual cues could be of any use in a species in which the solitary females are hidden from the males by the thick forest canopy.

The human female is also devoid of all external anatomical indices of impending ovulation, presumably as part of the process of suppression of periodic oestrus. However, a woman's

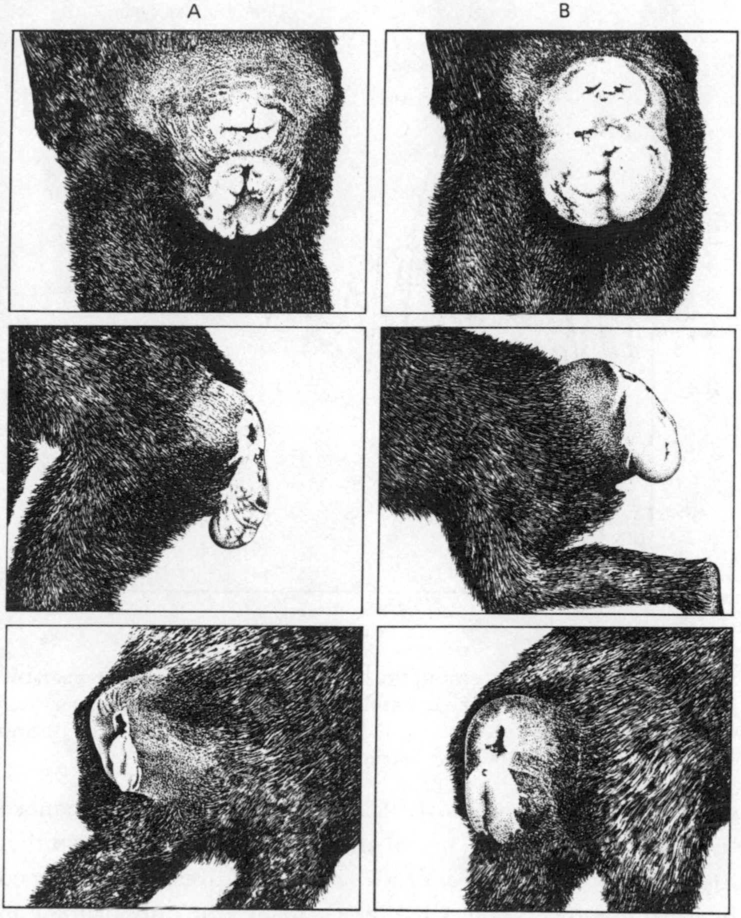

Fig. 1-12. The perineal region of female chimpanzees. A, at the time of menstruation, when they would not permit copulation, and B, at the time of ovulation, when they were in oestrus. (From C. E. Graham. In *The Chimpanzee*, vol. 3, p. 183. Ed. G. H. Bourne. Basel; Karger (1970).)

The origins of human sexuality

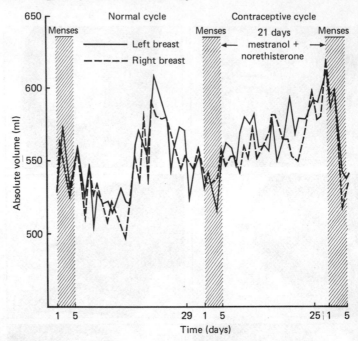

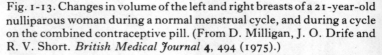

Fig. 1-13. Changes in volume of the left and right breasts of a 21-year-old nulliparous woman during a normal menstrual cycle, and during a cycle on the combined contraceptive pill. (From D. Milligan, J. O. Drife and R. V. Short. *British Medical Journal* **4**, 494 (1975).)

attractiveness to the opposite sex has seemingly been enhanced by breast development; social anthropologists have shown that the breasts are regarded as erotic in most cultures. As mentioned by Susu Ohno (Book 6, Chapter 1), we are the only primate in which a significant degree of mammary development occurs before the first pregnancy. Indeed, breast development is the first sign of impending puberty in the girl, and it makes evolutionary sense for her to become sexually attractive to the man at an early age, so that she can establish a pair-bond with the future father of her child before she becomes fertile.

This precocious breast development is due to an oestrogen-induced hypertrophy of the adipose and stromal tissue of the mammary gland, rather than to any development of the glandular epithelium. In fact, the size of the nulliparous breast gives no indication of its future lactational potential; this is determined by the degree of mammary enlargement that takes place during the course of pregnancy. The volume of the breast does change during the menstrual cycle (Fig. 1-13), but unlike the perineal tumescence of the chimpanzee, the maximal volume occurs immediately prior to menstruation, so it gives no clue as to the time of ovulation.

The nipples themselves are undoubtedly the most erogenous region of the whole breast; a two-point discrimination test shows that they have the highest tactile sensitivity, followed by the areolae, followed by the cutaneous tissue over the rest of the breast. There is no difference in breast sensitivity between prepubertal boys and girls, but after puberty there is a large increase in sensitivity in girls (Fig. 1-14). This sensitivity seems to fluctuate slightly, but predictably, during the course of the menstrual cycle. However, the greatest changes in breast sensitivity occur during pregnancy. During the latter weeks of gestation the nipples and areolae seem to be almost completely anaesthetized, and it is interesting that nipple stimulation at this time does not evoke a reflex discharge of prolactin. But within 24 hours of delivery, nipple and areolar sensitivity has returned, and nipple stimulation by suckling can now promote a reflex oxytocin and prolactin discharge. Whether this loss of breast sensitivity in late pregnancy is a central or a peripheral effect remains to be determined, but it is probably steroidally mediated.

The afferent stimuli from the nipple to the central nervous system are vital, not only for maintenance of the milk supply, but more importantly for the inhibition of hypothalamic activity and the suppression of ovulation. The nipple, in addition to being the umbilical cord of the newborn, is also the button to press if you want to ensure an adequate inter-birth interval.

The evolution of the human breast to become so oestrogen-sensitive seems to have carried with it certain disadvantages, since women are particularly prone to develop mammary cancer

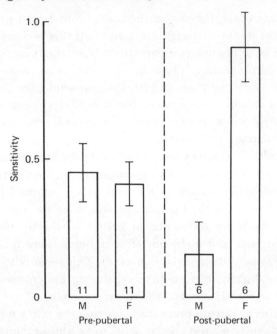

Fig. 1-14. Changes in tactile sensitivity of the cutaneous breast tissue in boys (M) and girls (F) before and after puberty, as assessed in a two-point discrimination test. The two points of a pair of geometric dividers, initially widely separated, were brought closer together until two distinct sensations could no longer be felt when touching the skin. The distance between the divider points was then recorded, and transformed mathematically so that increased sensitivity was reflected by an increased value. Numbers in columns indicate number of subjects tested. (From J. E. Robinson and R. V. Short. *British Medical Journal* 1, 1188 (1977).)

in later life; the incidence rate is about one woman in 20 in Europe and North America. Oestrogen is almost certainly involved in the genesis of the condition; the probability of getting breast cancer is increased by factors such as early age at menarche, late age at menopause, and late age at first pregnancy. By contrast, there has not been a single recorded case of mammary cancer in a chimpanzee, gorilla or orang-utan.

We cannot leave the topic of the female genitalia without some reference to the practice of female circumcision, still widely practised on young girls in Moslem countries. Circumcision is indeed a polite euphemism for the barbaric procedures that are performed, ranging from removal of the clitoris, to extirpation of the labia minora and majora, with suture of the vaginal orifice. Presumably the underlying concept was that if the woman could be denied the pleasure of orgasm, she would be less likely to be unfaithful to her lord and master. There can be no justification whatsoever for this cruel mutilation, although there seems to be a marked reluctance on the part of men in positions of authority to condemn the practice openly.

SEXUAL DIMORPHISM IN BODY SIZE

As pointed out in Book 6, Chapter 4, the presence or absence of sexual dimorphism in body size provides a valuable clue to the mating system that a species had adopted.

In monogamous species, where one male mates with only one female over an extended period of time, the forces of inter- and intrasexual selection balance one another out, so there is little dimorphism in body shape or size (Fig. 1-15). All the monogamous primates appear to conform to this rule, and apart from the differences in their external genitalia, it is extremely difficult to distinguish a male from a female, e.g. in marmoset or gibbon.

In polygynous species, where one male collects a harem of several females at one time, competition between the males is accentuated, leading to the development of increased body size and weapons of offence (Fig. 1–15). Once again there seem to be no primate exceptions and polygyny is indeed the commonest mating system. Thus in the macaques and baboons, the male is much bigger than the female. The male gorilla is about twice the size of the female (Fig. 5-4A), and shows strong development of the nuchal crest of the skull, and of his canine teeth which are certainly used in inter-male combat over possession of the harem of females. The male orang is also twice the size of the female, and shows the characteristic massive development of

29

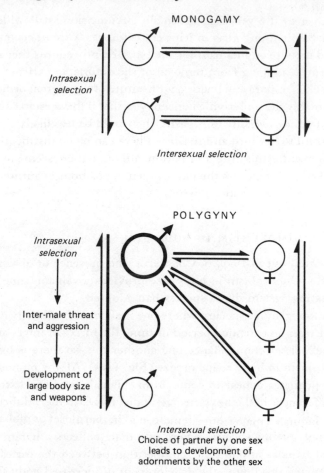

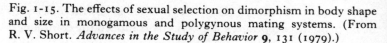

Fig. 1-15. The effects of sexual selection on dimorphism in body shape and size in monogamous and polygynous mating systems. (From R. V. Short. *Advances in the Study of Behavior* 9, 131 (1979).)

lateral facial folds, and enlarged canine teeth. As we mentioned earlier, the orang is a 'desocialized' ape, the males living a solitary existence in tree canopy, but the male will mate with any oestrous female that enters his range, which he will defend from the incursions of rival males.

One of the uncommonest types of primate mating system is

promiscuity, where several different males may take it in turns to mate with one female at one particular oestrus in a non-competitive manner. This is the state of affairs in chimpanzee communities, and since every male potentially has access to every female, and vice versa, one might imagine that as in true monogamy, the forces of inter- and intrasexual selection would balance one another out, leading to the absence of sexual dimorphism. Chimpanzees are indeed the least dimorphic of all the Great Apes, with the male being only about 10% heavier than the female, and showing no obvious sexual dimorphism in any facial characteristic, other than enlargement of the canine teeth. The persistence of this small degree of dimorphism may be accounted for by the fact that the males indulge in group defence of their territory against incursions from animals belonging to other communities.

So what can we begin to deduce about our own innate mating system from the nature and extent of our sexual dimorphisms? At any given age after puberty, men are on average appreciably heavier, and taller, than women. At the age of 30, for example, there is a 20 per cent difference in mean body weight between the sexes. Although this is not as pronounced as in the gorilla or orang, it is considerably more than in the chimpanzee. Men also have a much greater proportion of muscle per unit body weight than women, and in our naked, unshaven and unshorn state there are obviously marked facial differences between the sexes, although it is interesting that these do not extend to the teeth, which are no longer involved in aggression. All these facts taken together surely suggest that we are much more likely to be polygynous than monogamous or promiscuous. Indeed, Clellan Ford and Frank Beach, reviewing the patterns of sexual behaviour in 185 different human societies throughout the world, found that exclusive monogamy was practiced in only 16 per cent of them. In all the remainder, polygynous marriages were allowed, although economic considerations and the shortage of females (or excess of males) meant that in practice monogamous marriages were common. Serial monogamy may represent

a subtle compromise between on the one hand our basic polygamous natures, and on the other the need to maintain a long-term
male-female pair bond for the successful rearing of a highly
dependent infant that places severe constraints on the mother's
freedom of movement for an extended period of time. Although
the Christian ethic has been one of lifelong monogamy, the rising
divorce rate in the Western world suggests that when cultural
restraints are relaxed, we may revert to type.

Leonardo da Vinci wrote: 'The act of coitus and the parts
employed therein are so repulsive, that were it not for the
beauty of the faces and the adornments of the actors and frenetic
state of mind, Nature would lose the human species'. In this he
expressed many peoples' inherent distaste of pornography. But
we cannot shut our minds off from sexuality, because it is our
very stuff and being, and constantly intrudes upon our daily
lives. E. O. Wilson of Harvard has recently written: 'The most
distinctive feature of the sexual bond, one of overriding significance for human social organization, is that it transcends sexual
activity...Human beings are connoisseurs of sexual pleasure.
They indulge themselves by casual inspection of potential
partners, by fantasy, poetry, and song, and in every delightful
nuance of flirtation leading to foreplay and coition. This has little
if anything to do with reproduction. It has everything to do with
bonding...Love and sex do indeed go together'.

Because sexuality is such a very basic part of all of us, maybe
it has become a natural focal point for all social organizations that
seek to dominate the mind of man. By proscribing as sinful that
which is innate and natural, man is constantly reminded of his
frailty and unfaithfulness, just as the bit in the stallion's mouth
is a constant reminder of his master's supremacy. Have the
forces that moulded the sexual behaviour of our lonely hunter-
gatherer ancestors over millions of years of evolution endowed
us with adequate sexual restraints to cope with our overcrowded
existence in the twentieth century? Or must we depend on
additional, man-made laws which are given the stamp of

authority by being sanctified in the name of God? Because we are involved in mankind, we cannot know the answer; we can only attempt to discover it by painful trial and error. And one is reminded of the old adage 'Higamus, Hogamus, woman is monogamous. But Hogamus, Higamus, man is polygamous'.

SUGGESTED FURTHER READING

The fate of the foreskin. A study of circumcision. D. Gairdner. *British Medical Journal* ii, 1433 (1949).
Penis sheaths: a comparative study. P. J. Ucko. *Proceedings of the Royal Anthropological Institute* **27** (1969).
The evolution of human reproduction. R. V. Short. *Proceedings of the Royal Society of London, Series B* **195**, 3 (1976).
Time of onset of sperm production in boys. D. W. Richardson and R. V. Short. *Journal of Biosocial Science* Supplement 5, 15 (1978).
Humans and apes are genetically very similar. E. J. Bruce and F. J. Ayala. *Nature, London* **276**, 264 (1978).
Sexual Selection and its component parts, Somatic and Genital Selection, as illustrated by Man and the Great Apes. R. V. Short. *Advances in the Study of Behavior* **9**, 131 (1979).
Sexual dimorphism, socionomic sex ratio and body weight in primates. T. H. Clutton-Brock, P. H. Harvey & B. Rudder. *Nature, London* **269**, 797 (1977).
The Original Australians. A. A. Abbie. Sydney; Rigby (1969).
The Erotic Arts. P. Webb. London; Secker & Warburg (1975).
On Human Nature. E. O. Wilson. Cambridge, Mass.; Harvard University Press (1978).

2 Human sexual behaviour

John Bancroft

Until the last few years, research into human sexuality has been in the hands of a dedicated few. Robert Dickinson, Alfred Kinsey, William Masters and Virginia Johnson, in particular, have each contributed to our understanding and knowledge to an extent that is rare in any scientific field and yet, because of their subject matter, their names have remained outside the mainstream of science. Circumstances are now changing, sex research is becoming increasingly widespread and respectable, and at the same time is being subjected to more normal scientific scrutiny. As a consequence we are repeatedly required to challenge previously held assumptions. A re-appraisal of many basic concepts is necessary and we will be considering some of these in this chapter.

Human sexual behaviour serves a variety of functions in addition to reproduction and the simple provision of pleasure. Sex can have a binding effect on inter-personal relationships, be used to maintain or bolster our self-esteem, to exert control or dominance in our personal relationships, as well as to express hostility; it can be used also for material gain. With such a variety of purposes, we should not be surprised to find that the determinants of our sexual behaviour are varied and their interactions complex. In this chapter we will concentrate on some of the more biological aspects. The enormous importance of social and inter-personal factors in shaping our sexual behaviour is acknowledged, but they will only be considered briefly here as they interact with the biological mechanisms.

Fundamental to sexual behaviour is a set of physiological responses that I will call the 'sexual response system'. They occur as we respond to sexual stimulation and include that uniquely sexual phenomenon, the orgasm. We will look at this

34

system in some detail. Also fundamental is the potential for this system to be activated or for us to seek to activate it. This aspect is often referred to as sexual drive or 'libido'. I prefer to call it 'sexual appetite'. It demands careful consideration as it relates to many of the outstanding questions about human sexuality, and it will need to be conceptualized in terms that permit its scientific investigation if those questions are to be answered.

Our 'sexual response system' may be intact and our sexual appetite strong but what are the stimuli that we respond to? What type of person or activity has this effect? How do we establish our 'sexual preferences'? As we shall see, it is in this area that our ignorance is most profound. We will consider these three aspects in turn: the sexual response system, sexual appetite and sexual preferences.

THE SEXUAL RESPONSE SYSTEM

What happens when we are stimulated to respond sexually? The most predictable and often the earliest changes occur in the genitalia and are principally vasocongestive in nature. In the man, this leads to erection of the penis and in the woman to a 'sweating' response from the wall of the vagina which subsequently thickens and lengthens as do the labia as they become congested with blood. These responses have an obvious function. The 'sweating', probably due to a simple transudate from the congested vaginal vessels, serves to lubricate the vagina which in its altered state is well suited to receive the fully erect penis (Fig. 2-1).

Erection of the penis is subjectively an obvious response; the vaginal changes are less so but their timing is usually similar. A change in the pattern of blood flow in both the penis and the vagina may be measurable within 10 to 30 seconds of the onset of any form of erotic stimulation, whether tactile or mediated by the brain (Figs. 2-2 and 2-3). The precise vascular mechanisms involved are not as yet fully understood. In the penis there is

35

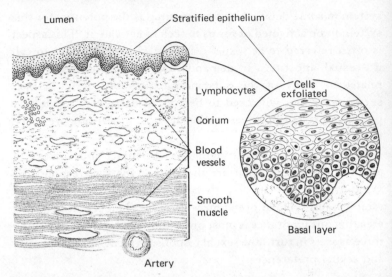

Fig. 2-1. The vaginal wall. Lined with stratified epithelium, the wall is rich in blood vessels. During response to sexual stimulation these vessels become congested and fluid passes from them between the epithelial cells to the vaginal lumen. This transudate provides the basis for vaginal lubrication, probably aided by breakdown products of the exfoliated epithelial cells. (From G. Wagner and R. J. Levin, 'Vaginal fluid'. In *The Human Vagina*. Ed. E. S. E. Hafez & C. Evans. Amsterdam; Elsevier (1977).)

thought to be a rapid increase of blood flowing into the erectile tissues (i.e. the corpora cavernosa and corpus spongiosum) following the dilatation of localized arteriolar valves or 'bolsters' (Fig. 2-4). In addition, the venous outflow from the erectile tissues is probably reduced by other valvular mechanisms. The combination of these two processes leads to a rapid swelling and stiffening of the penis, though it is not clear whether this is sufficient to account for the high pressures within the corpora cavernosa necessary for a full erection (Fig. 2-5). In some animals increased tone in muscles surrounding the erectile tissues is probably essential, but the relevance of such a mechanism in man is not known.

Even less is known about the vascular changes in the vaginal wall, though it is possible that similar mechanisms are involved.

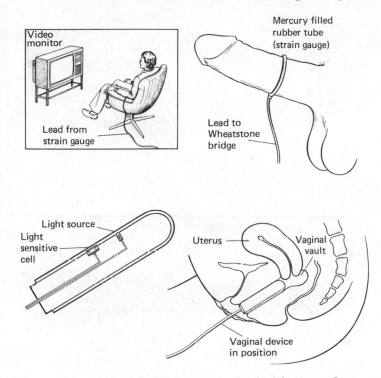

Fig. 2-2. Measurement of genital responses in the laboratory. In men, increases in penile size can easily be measured by a simple mercury-in-rubber strain gauge (as illustrated) or by other forms of strain gauge. In women, the easiest method involves the measurement of reflected light from the vaginal wall. The amount of light reflected depends on the amount of blood in the tissues. A device, similar in shape to a tampon, carries the light-sensitive cell and light source, and is inserted into the vagina.

Mechanical requirements are obviously different, and the extent of the vasocongestion is more widespread, involving not only the vagina and labia but also the uterus and the utero-vaginal supportive tissues. (In the male there are further but less extensive vasocongestive changes in the scrotal wall and the testes which enlarge in size.) Other changes in the genitalia of both male and female do occur that are not apparently vascular in origin. There is dilatation of the inner third of the vaginal

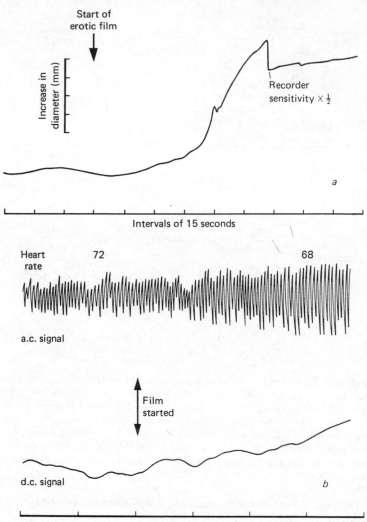

Fig. 2-3. *a* Penile response in the male. Erection starts to develop about 30 seconds after onset of the erotic film, reaching almost full erection about 60 seconds later. The mercury-in-rubber strain gauge measures changes in circumference of the penis. *b* Vaginal response in the female. Vaginal response to an erotic film, is as rapid as the penile response in the male. The two signals indicate different physiological changes; the AC signal shows pulse amplitude in the vaginal wall, whilst the DC signal is related to the amount of blood contained in the vaginal wall.

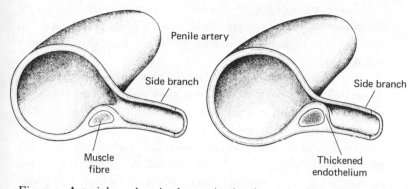

Penile artery

Side branch

Side branch

Muscle
fibre

Thickened
endothelium

Fig. 2-4. Arteriolar valves in the penis. A schematic representation of the valves thought to control blood flow into the erectile tissue. They allow side branches to be either more or less occluded or completely patent, leading to a sudden increase in blood flow. Similar valvular mechanisms may control venous drainage from the erectile tissue and have also been described in the erectile tissue of the labia and vestibule in women. (Based on R. L. Dickinson. *Atlas of Human Sexual Anatomy.* Baltimore; Williams & Wilkins (1940).)

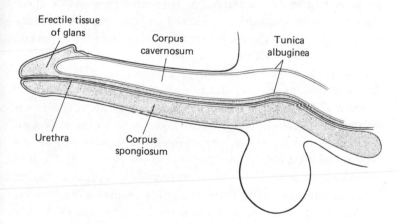

Erectile tissue
of glans

Corpus
cavernosum

Tunica
albuginea

Urethra

Corpus
spongiosum

Fig. 2-5. Erectile tissue of the penis. The corpus cavernosum provides the main support for the erect penis. High pressures develop contained within the thick fibrous tunica albuginea. The glans and corpus spongiosum also show tumescence but not to the same extent. In the condition of priapism, irreversible erection affects only the corpora cavernosa, and surgical treatment can involve an anastomosis between the cavernosum and glans or spongiosum.

barrel, elevation of the uterus in the woman and elevation of the testes in the man, in both cases owing presumably to smooth muscle contraction and both of uncertain functional significance.

Not only are these genital changes likely to be the earliest to occur following sexual stimulation, they may also occur unassociated with any other evidence of cardio-vascular or general somatic change (Fig. 2-6). This may be in response to local tactile stimulation in which vascular reflexes involving the lower part of the spinal cord may be implicated. More striking is their occurrence in response to visual, auditory or just imaginary stimuli whose effects must be mediated by the brain. We therefore know that these genital, vascular responses are directly controllable by nerve pathways from the brain and are not necessarily part of a more generalized cardio-vascular response. In addition to these structural changes, there is an increase in the tactile sensitivity of the genitalia and other nearby erogenous areas leading to an escalation of response. Presumably these changes are in part due to altered central control, but the anatomical changes themselves may contribute to the enhanced sensitivity by their effects on sensory nerve endings.

Clearly, physiological changes other than those in the genitalia do occur. What are they, and how do they relate to the genital responses? It is appropriate at this stage to consider the concept of 'sexual arousal'. This is usually taken to mean the collection of physiological and subjective changes that occur during sexual response. Alfred Kinsey and his colleagues, perhaps the first to grapple seriously with this concept, considered sexual arousal to be a continuum ranging from negligible arousal to high levels of sexual excitement preceding or accompanying orgasm. A wide variety of physiological responses, including more general cardio-vascular and myotonic changes as well as the genital, were represented on this continuum, each increasing in intensity as arousal increased. This 'unitary' concept of sexual arousal reflected the ideas of 'general arousal' and 'activation' that were prevalent at that time. The unitary concept of general

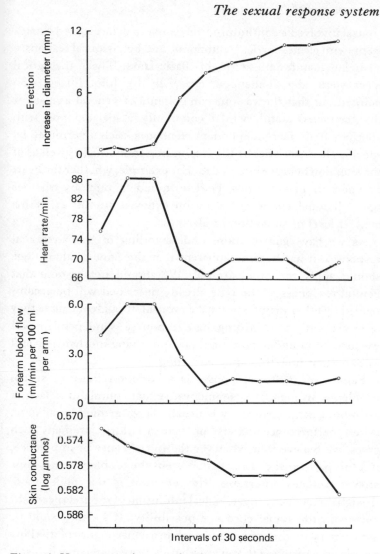

Fig. 2-6. How an erection can develop accompanied by a decrease in other indices of autonomic arousal. In this normal male, there was an initial arousal response seen in heart rate, forearm blood flow and skin conductance, followed by a decline as erection developed. (From J. Bancroft and A. Mathews, 'Autonomic correlates of penile erection'. *Journal of Psychosomatic Research* **15**, 159–67 (1971).)

arousal involved a continuum from coma to disorganized excitement, with electrocortical, autonomic and behavioural responses all being manifestations of this basic state. These theoretical ideas were later challenged, especially by John Lacey, who pointed out that the various components of general arousal or the 'activated state', whilst commonly associated one with another, are in fact independent responses, each determined by specific characteristics of the responding individual as well as of the situation being responded to. For example, whether the heart rate goes up or down may be determined as much by whether the individual is concentrating on external stimuli as by his general level of activation or alertness.

As we have gained more understanding of human sexual responses, it has become apparent that the same qualifications should apply to sexual arousal. We should not assume that genital responses of the type already described will be accompanied by other manifestations of arousal, though of course they often are. Only by regarding these responses as independent will we succeed in understanding their underlying mechanisms and in particular how they may go wrong.

Failure of genital responses is a common cause of sexual problems. In men, the commonest sexual complaint is lack of erection. Such a deficiency is usually accompanied by anxiety, called 'performance anxiety' or 'fear of failure'. In many such cases, we believe that whatever the initial cause of impotence, it is this performance anxiety that keeps the problem going. How anxiety actually interferes with erection is not understood, though direct neurophysiological inhibition of the erectile mechanisms in the spinal cord is a possibility. If so, this could be associated with the anxiety rather than a consequence of it. Many people have assumed that sympathetically mediated mechanisms are responsible for erectile failure and that anxiety activates the sympathetic nervous system; there is no evidence to support this rather naive explanation.

Some cases of erectile failure, especially in the older male, are probably caused by local pathological changes such as degeneration of the arterial mechanisms or failure of neural mediation through peripheral neuropathy or injury. Diabetic males commonly suffer from erectile failure and both vascular or neural pathology could be involved. The prevalence of such peripheral causes is not yet known. Attempts are being made to distinguish between local 'organic' and psychological causes by measurement during sleep. The occurrence of normal nocturnal erections is considered evidence of a psychological cause. Although this approach holds promise, its value has not yet been demonstrated.

In the woman, failure of genital response results in a vagina that is anatomically unprepared and unlubricated. Also absence of the normal ballooning of the vaginal vault may result in the cervix being more susceptible to buffeting during intercourse. Although the vaginal response is less crucial than penile erection for coitus to take place, entry of the penis into the dry unprepared vagina is often painful. This will further inhibit any sexual response or enjoyment on the woman's part so that a vicious circle becomes established, with performance anxiety or fear of pain causing further havoc.

As with the male, the initial failure may be psychologically determined or due to local mechanisms. Vaginal infections are a common cause and possibly oral contraceptives may also produce this effect in some young women. In the post-menopausal woman, oestrogen deficiency, if severe enough, will result in a dry, unresponsive vagina.

ORGASM

Arousal is often regarded as a psychological state that facilitates action. The activity that accompanies and is perhaps facilitated by sexual arousal may culminate in orgasm. This response then imposes its own physiological changes which also need to be regarded separately. But whereas there is considerable variation in the pattern and intensity of bodily changes when responding to sexual stimuli, orgasm, when it occurs, is more predictably

Human sexual behaviour

preceded by autonomic arousal – blood pressure, heart rate and respiratory rate increase. This predictable pre-orgasmic phase may be relatively brief and may be an integral part of the orgasmic response. The degree of arousal in the earlier non-orgasmic stages of sexual response varies not only between individuals but for the same individual from occasion to occasion. At least part of this variability must reflect aspects of the situation that are not specifically sexual but are nevertheless arousing, e.g. novelty or risk. Thus a man may enjoy sexual contact with his wife with full and satisfying genital responses but relatively modest non-genital arousal, at least until orgasm is imminent. The same man embarking on a sexual encounter with another woman for the first time may show extreme central and autonomic arousal even before his genital responses are established (this point is reinforced by the finding that sudden death during sexual intercourse is reported much more commonly in extra-marital than in marital encounters).

Orgasm is the most mysterious and probably the most specific of the sexual responses. Its subjective qualities are notoriously difficult to define, perhaps because our perception is often impaired or at least altered at this time. The usual descriptions involve a rapid and dramatic increase in intensity of sensations and 'tension' followed by a rather sudden release of tension to a state of calm. It is sometimes likened to a sneeze but obviously with rather different qualities of sensation involved. These subjective experiences are accompanied by measurable physiological changes. The cardio-vascular and myotonic activity that immediately precedes the orgasm is rapidly reversed following it. The climax of the experience is typically associated with rhythmic muscular contractions – in the female, especially in the pelvic muscles surrounding the vagina, and in the male, in the ischiocavernosus and bulbospongiosus muscles of the penis. Clonic and tonic contractions of general body musculature may also occur in both sexes.

One can only speculate about the neurophysiological processes underlying orgasm. It is as if there is a build-up of central excitation in specific centres in the central nervous system, to a point where a neuroelectrical discharge is triggered, resulting in a spread of excitation through adjoining parts of the central nervous system. The extent of spread of the excitation may depend not only on the constitutional characteristics of the individual central nervous system, but also on the level of 'central excitatory state' that had developed prior to the orgasm being triggered. Such a model poses more questions than it answers, but is at least in keeping with other characteristics of the central nervous system.

So far, my description of the orgasm applies equally well to men and women. But there are some important and theoretically puzzling sex differences. Orgasm in the male is usually accompanied by a discharge of semen. This ejecting process is initiated shortly before the orgasm is experienced and the rhythmic muscular contractions occur, but once initiated the process is virtually unstoppable. Hence the characteristically male experience of 'ejaculatory inevitability'. The ejaculatory quality of seminal discharge, the spurting or pumping ejection of semen from the urethra, is at least in part dependent on the rhythmic contractions in the muscles of the penis already mentioned. Without that pumping action, the semen is simply voided, a process called emission rather than ejaculation. Once again these two components may be dissociated, as in some men with severe 'premature ejaculation', when there is no appreciable orgasm experienced and consequently no muscular contractions. The semen simply seeps away in what should be called premature emission. Orgasm may also occur without any emission or ejaculation, the so called 'dry run' orgasm. This, of course, is normal when orgasm occurs in the pre-pubertal male but can occur in some adults, either spontaneously or as a consequence of drug effects. It raises the possibility that seminal emission is not an integral part of orgasm though the two are normally activated at the same time. There is no female counterpart of seminal emission.

A further sex difference lies in the immediate consequences

of orgasm. In the male, orgasm is typically followed by a refractory period during which further erotic stimulation, either centrally or peripherally mediated, is totally ineffective. There is then a gradual return to normal responsiveness. Such a 'refractory' period is said not to occur in the female though this is still in dispute. But there is general agreement that a reasonable proportion of women, if not all, are capable of experiencing orgasm without any significant refractory period and of continuing to respond with repeated or multiple orgasms in a way that must be very unusual in the adult male.

The nature of the refractory period and its intriguing sex difference takes us to the next fundamental question about the orgasm: where in the central nervous system does it occur? Is it a spinal cord mechanism, a neuroelectrical event in the limbic system of the brain or a combination of both? There is little doubt that, as with non-genital arousal, the intensity of this phenomenon varies enormously. For some individuals, orgasm is a relatively trivial increase in preceding sexual sensations followed by a decline in tension and in responsiveness, and a phenomenon like that could easily be understood as of local spinal origin with minimal spread through the spinal cord. With other individuals the event may be accompanied by alterations of consciousness, bordering on the unconscious and a striking loss of motor control. Such events have led to the orgasm being likened to an epileptic fit, and then it is difficult to avoid the conclusion that some major neurophysiological event within the brain has occurred. The very limited human neurophysiological data available, together with some relevant non-human primate evidence, would support this view, suggesting that electrical discharge from the septum, fornix and hippocampus underlie the phenomenon. It is possible that the neuroelectrical event combines both excitatory and inhibitory discharges. The first would be responsible for the intensity of the orgasmic experience, the latter, presumed to be similar to the hippocampal discharges observed by Paul Maclean and others in monkeys, being responsible for the post-orgasmic refractory state. This variable combination of excitatory and inhibitory mechanisms could explain many of the otherwise obscure aspects of sexual response. Those cases of severe premature emission, previously mentioned, which occur with negligible orgasm and often with no erection, are typically followed by a prolonged and severe refractory period. Perhaps in such cases the inhibitory discharge has predominated.

An explanation for the sex difference in refractoriness is still wanting. Sex differences in the effects of cerebral amines on sexual behaviour have been described in some lower animals and this may give a clue. But one is also tempted to wonder whether the peculiarly male phenomenon, the seminal emission, may have something to do with it. This is made less likely by the findings of Frank Beach and his colleagues in rodents. After ejaculation had been induced by passing electric shocks through the spinal cord, presumably eliciting a spinal response, the refractoriness of the male's sexual behaviour was less evident than after ejaculation occurring during normal coitus. Comparable data from the human male or from other primates would be interesting.

Orgasm is not a response that is easy to investigate in man or even in the experimental animal. Some recent developments in non-intrusive recordings of electrocortical activity may however throw further light on the problem.

Just as men and women differ in interesting ways in their normal orgasms, so do they with their 'orgasmic problems'. It is commonplace for a man to ejaculate too quickly, and learning control over ejaculation is very much a part of normal sexual development for the adult male. Performance anxiety, however, definitely aggravates this control problem and hence established premature ejaculation is common. The woman, by contrast, is often older before she realizes her full orgasmic potential. So, particularly nowadays when the female orgasm is widely discussed, women commonly worry about not being orgasmic or taking too long. In this case the effect of performance anxiety is, ironically, the opposite – it delays or inhibits the orgasm even more. One of the commonest sexual problems of young couples is this dual orgasmic difficulty, the man 'coming' too quickly, the woman too slowly or not at all, each problem aggravating the other.

A much smaller proportion of men complain of delayed or absent orgasm and ejaculation. They show a pattern very similar to that of the anorgasmic women. Such problems seem likely to be psychogenic though conclusive evidence is still lacking. It is very unusual for a woman to complain of premature orgasm.

SEXUAL APPETITE

Our sexual responses, once activated, become to some extent self-reinforcing and hence tend to continue to orgasm or some other terminating event. But what induces us to activate the system in the first place? Why do some people seek sexual outlets frequently, others rarely or not at all? What do we mean by high or low sexual drive, libido or appetite?

The concept of drive to explain directed behaviour has come in for much criticism. R. S. Peters, whilst pointing out many of the logical flaws in 'drive theories' of behaviour, has also made the important point that the concept of drive – say sexual drive – implies some internal state of the organism with an impelling quality. This is a further example of a 'unitary' concept that discourages the search for a possible variety of underlying mechanisms. A tendency to look for the simple, all-embracing 'whole' may serve to obscure what would be an obvious 'part'. It is reasonable to start with the premise that our sexual behaviour results from an interaction between some internal state or states of the organism and some external stimulation. Let us consider 'appetite for food' as a means of clarifying the distinction. Various biochemical states, such as hypoglycaemia, and physiological responses, such as stomach contractions, may constitute an internal state, which in the appropriate circumstances will be experienced as hunger for food. For example, the smell or sight of food may result in the recognition 'that smells good – I feel hungry'. Alternatively, without any external food stimulus, circumstances may suggest that 'it is time to eat'. An appointment that normally finishes at lunchtime has come to an end – the physiological changes occurring are quickly interpreted as hunger. In other circumstances, however, if some unexpected distraction arises, attention to other matters may result in these physiological states being unnoticed or re-labelled as tension, abdominal discomfort or irritability. Thus 'appetite for food' is how we experience this interaction between internal, physiological states and external stimuli, reflecting the effects of both previous learning and current cognition. 'Appetite' is a good term because it emphasizes quality of experience rather than any assumed underlying mechanism.

Such an analysis has proved to be necessary in the substantial efforts that have been made to elucidate the determinants of eating behaviour. The same care and effort has not yet been taken in attempts to understand sexual behaviour. The simplistic notion has prevailed of 'sexual drive' as admittedly being susceptible to psychological inhibition, but basically endowing the organism with an energy impelling it towards sexual activity. In principle, the behaviour is seen as an instinct. The theoretical ideas of sociologists John Gagnon and William Simon have provided an interesting reaction to this 'instinctive' model. The level or intensity of our sexual activity is in their view determined by the 'sexual script' that society encourages us to adopt. Thus one individual may learn to be 'highly sexed' and another 'sexless' – the difference in action potential, it is suggested, has little to do with any biological substrate *per se*. Whilst such ideas are useful in drawing attention to some other potentially important determinants in the total picture, they predictably go too far in rejecting biological factors, whilst at the same time failing to suggest how the ideas may be verified.

But to further our understanding of human sexual behaviour and its problems, we must try to define the relevant characteristics of the 'internal state'.

Human sexual behaviour

The closest to an operational definition of 'sexual drive' has been 'sexual arousability' as defined by Richard Whalen. This is an individual's capacity to respond to erotic stimulation with an increase in sexual arousal (i.e. with changes in the sexual response system as described above). This has the merit of being open to experimental investigation, though the standardization of sexual stimuli for such purposes does pose problems. In these terms, sexual drive can be quantified as a stimulus threshold for sexual arousal to occur. Someone with 'high sexual drive' will find his sexual response system activated more easily and vice versa. But this is not an explanatory definition and it does not attempt to identify any possible mechanisms involved. Is 'low arousability' in general comparable to that particular form of it which occurs during and following the post-orgasmic refractory period when there is apparently an active inhibition of responsiveness to all kinds of stimuli? Does the level of arousability apply whatever the type of stimulus involved? Or are there different factors affecting sensitivity to tactile stimuli, visual stimuli, etc? Does 'arousal' depend on the spontaneous occurrence of an external stimulus or is there a further factor that leads the individual to seek out erotic stimuli? What part does imagery play in this respect? Fantasies can be powerfully erotic in the absence of an external stimulus, but the production of those fantasies is an active process. What is required for that to happen in the first place? Andrews has suggested that 'persistence of attention', which has been shown to be influenced by androgen levels in some birds, could increase the likelihood of sexual stimuli being effective. This might apply particularly to images that require active attention for their maintenance. Could there be other cognitive mechanisms involved? At this stage we can only speculate, though it would seem appropriate to look for several relevant mechanisms, each of which would be differentially susceptible to hormonal, biochemical or environmental influences. It is worth remembering in this respect that 'sexual appetite', like 'appetite for food', may be one of several factors adversely affected by general metabolic disease. Thus more normal variations of metabolic rate, and the effects of fatigue, may contribute to the general picture.

Until recently, little attention had been paid to 'loss of sexual appetite' as a sexual problem. Now, this is not only being encountered frequently in both men and women, but also presents some of the most difficult therapeutic problems. Not infrequently, this 'low appetite' does not preclude normal sexual responses if the appropriate stimulation is forthcoming. When it is the woman who is so affected, the problem may be relatively minor, but when it is the man, with a partner who expects him to take the initiative, then this can cause serious difficulties.

THE ROLE OF HORMONES IN HUMAN SEXUALITY

The study of animal sexuality has led to two generally accepted conclusions. First, a proportion of the variance in levels of sexual activity amongst animals is genetically determined. Secondly, hormones play a fundamental part in controlling sexual activity in both male and female. These two issues are probably connected, as animal sensitivity to hormone control is in part influenced by genetic factors. For the human subject, we can acknowledge the likely importance of genetic factors whilst accepting that it is extremely difficult to evaluate them. With hormones, we are on slightly easier ground. We commonly attribute a large part of our potential for sexual response to hormones, or at least assume that hormones are necessary for a normal sexual appetite. At the same time, we accept that the relative importance of hormonal control, in comparison with other factors such as social learning, declines as one progresses up the animal kingdom, so that in man it becomes easily obscured if not made redundant.

In non-primate male animals, androgens appear to be necessary for normal sexual activity. After castration sexual activity declines, though the extent and timing of this varies considerably. Some animals continue with a low level of sexual activity indefinitely. Giving androgens (in particular testosterone) usually restores their sexual behaviour to its pre-castration level. In-geborg Ward has suggested that androgens maintain sexual function in the male animal at three levels – the normal tactile sensitivity of the penis (mainly due to penile spines which are present in most mammals but not man), integrity of the spinal reflexes and normal responsiveness of the central nervous system.

In the human male, evidence of the effects of castration or hypogonadism has been largely anecdotal until recently, though it has been consistent with the general findings in other animals. There is usually a predictable decline in sexual activity but again of variable timing and intensity. Recent controlled studies have confirmed these impressions, indicating that the decline usually becomes apparent two to four weeks after withdrawal of androgen supply. Conversely, the restoration of androgen effects occurs somewhat quicker (within a week or two). So far the most predictable early effect has been in sexual interest or appetite. Ejaculation tends to disappear fairly soon after this, whereas impairment of erectile function is much less predictable. In so far as sexual activity depends on initiative by the male, then the frequency of this activity will decline. But if the female partner takes the initiative the sexual responses may occur, albeit a little sluggishly. Some limited laboratory evidence suggests that erec-tion to strong visual erotic stimuli (i.e. films of explicit sexual activity) continue in these androgen-depleted men, even though sexual interest has largely disappeared and activity declined. Erections in response to fantasy on the other hand may be more dependent on androgens (Fig. 2-7). This pattern of change was also observed in a control study of the effects of an anti-androgen (cyproterone acetate) where, in addition to sexual interest and activity, erections to fantasy were suppressed, whilst erections to erotic films continued more or less unaffected.

So far this evidence is confined to relatively short-term androgen withdrawal (usually two to four months) and different effects may emerge over long periods. But, in the short term, it looks as though androgen withdrawal affects sexual 'appetite'

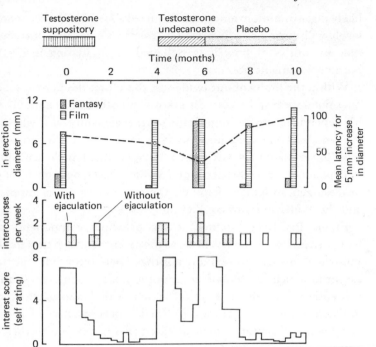

Fig. 2-7. Effects of androgen withdrawal and replacement on sexual interest, activity and erectile response to erotic stimuli in a hypogonadal male. This man was originally on testosterone suppositories (see top line of graph). These were stopped for two and a half months. He was then started on testosterone undecanoate (160 mg daily) for two months. This was then changed to placebo which continued for a further four months (the comparison was carried out 'double blind'). His erections to erotic films carried on throughout the experiment, though their latency did vary. The erections to erotic fantasy in contrast appeared to be androgen dependent, particularly to fantasy at the beginning of the session. Sexual intercourse stopped during androgen withdrawal and declined during placebo, with ejaculation failing first. Sexual interest, with self-rating, showed obvious androgen dependence.

(possibly by affecting cognition as involved in fantasy production) and ejaculation. Erections, at least to certain types of stimuli, appear to be less predictably affected. We can say nothing yet about the effects of androgen on tactile sensitivity of the penis in men. We can also say very little about the amount of androgen necessary for normal function in any of these respects. It seems

53

likely that a minimum amount is required above which increasing levels will make no difference (at least to sexuality). But what that amount is or how it is reflected in circulating levels of hormones remains obscure.

Within the limits of our evidence, therefore, the human male does not appear to be very different from other male animals in his dependence on androgens for normal sexuality. But what of the female? Here the animal story is far less clear. Whereas in most lower mammals, such as rodents, normal female sexuality is dependent on oestrogens or in some cases oestrogen plus progestogen, in female primates the situation is more complex and the evidence more conflicting.

Frank Beach has defined three principal components of female sexuality: attractiveness, as the extent to which the female motivates the male to sexually approach her; receptivity, as the extent to which the female accepts the male's approaches; and proceptivity, as the extent to which sexual approaches are initiated by the female (Fig. 2-8). Joe Herbert and his colleagues (see Book 4, Chapter 2) have stressed that in the female rhesus monkey androgens (testosterone and/or androstenedione) are necessary for normal proceptive behaviour. Richard Michael and others have disputed these findings, claiming that such effects of androgen are either pharmacological (i.e. dependent on dose levels that do not occur physiologically) or due to aromatization of androgens to oestrogens. Attractiveness, at least in some respects, is generally agreed to be oestrogen-dependent in the female rhesus, and oestrogens may be necessary, amongst other things, for normal proceptivity and receptivity.

In the human female, one of the most obvious sources of evidence is the study of sexual behaviour during the menstrual cycle. Although a number of such studies have been reported, the results have been conflicting and inconclusive. At best, they suggest that some women show a peak of sexual activity around the middle of the cycle when androgen and oestrogen levels are relatively high, but the majority of women apparently do not. Other women report peaks of activity either just before or just

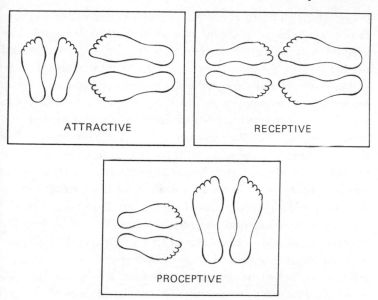

Fig. 2-8. The components of female sexuality described by Beach can be applied to the human subject as illustrated above. The attractive female is approached by the male; the receptive female responds to his approach; the proceptive female approaches first. But human interaction is often more complex than this. If the female emits subtle cues that encourage the male to make an approach is she being proceptive? Once sexual activity starts, the behaviour of each partner may be much less 'sex stereotyped' than is the case with lower animals.

after menstruation. In many women, there is no discernible cyclical pattern. Part of the problem stems from the fact that the human female as well as the female of other species, does not typically take the initiative in sexual interaction. Although there is obvious reciprocity – the male responding to encouraging but possibly discreet cues from the female before making any unequivocal approach – sexual activity is nevertheless linked more obviously to the male's initiative. Thus it has been suggested that mid-cycle peaks of activity stem from a hormonally induced increase in sexual attractiveness of the female to her mate. Conversely, a decline in sexual activity sometimes observed in the luteal phase has been attributed by some workers to progesterone-induced reduction in the woman's attractiveness rather than in her own sexual interest. Obviously, we would like to know how women feel when they are not with their sexual partners. Marie Stopes was perhaps the first to comment on this. In her book *Married Love* she described a pattern commonly reported by such women in which increased sexual desire was experienced at approximately fortnightly intervals, one peak just before menstruation, the other about a week after it ceased. This second peak would coincide with fairly high follicular levels of oestrogen or androgen. A recent study attempted to separate the woman's spontaneous interest from that of her partner, and it claimed to find a definite peak of female interest around ovulation, except in those using oral contraceptives who do neither ovulate nor show the normal cyclical changes in hormone levels (Fig. 2-9). A weakness of this study is that ovulation was estimated from the time of menstruation rather than from any hormonal indicator. Further confirmation of this association is needed, with hormonal measurements as well as indicators of sexual interest that are clearly independent of the male partner. Measurement of genital response to erotic stimuli in the laboratory is one obvious method of achieving this. The direct comparison of such measures with circulating levels of steroid hormones may help to establish whether such associations, when they exist, are with oestrogen or androgen, or a combination of the two, or even a negative association with progesterone. In those women who report peaks of interest either just before or shortly after menstruation, it is difficult to exclude the simple effect of anticipation or recent abstinence during menstruation. Again, sufficiently careful enquiry should enable us to resolve the problem.

Surprisingly, we have virtually no evidence of the levels of either androgens or oestrogens in sexually unresponsive women as compared with normals. For some time we have known that the administration of exogenous androgens enhances sexual interest and response in some women. Only recently has an adequately controlled study supported this observation but we do not yet know whether such effects are pharmacological or

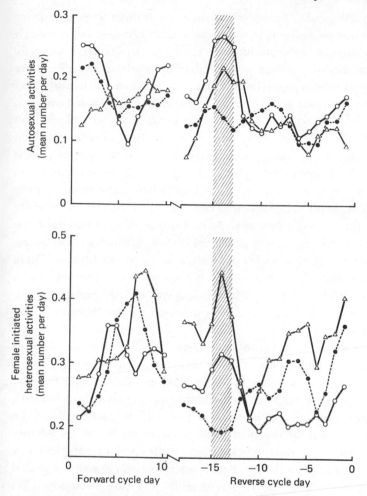

Fig. 2-9. The study shows that sexual activity initiated by the human female, or confined to her (e.g. masturbation), shows a peak around ovulation (shaded area). This peak does not occur in women on oral contraceptives (pill subjects) who do not ovulate and do not show hormone peaks at this stage. ●, Pill subjects; ○, intrusive non-pill subjects (sheath, diaphragm); △, non-intrusive, non-pill subjects (intrauterine device, vasectomy). (From D. B. Adams, A. R. Gold and A. D. Burt. 'Rise in female sexual activity at ovulation blocked by oral contraceptives'. *New England Journal of Medicine* **299**, 1145–50 (1978).)

physiological. The role of androgens in human female sexuality remains an enigma. It is probable that, given in large enough amounts, they will produce effects the woman would not normally show because of her low circulating levels. Such effects are growth of muscle and body hair, change of voice and hypertrophy and increased sensitivity of the clitoris. The last-named reaction, for local rather than central reasons, may have an influence on the woman's sexuality. But what is the relevance of such masculinizing changes for normal female sexuality? If androgens do play a fundamental part in normal female sexuality, they do so at a level that would be quite insufficient in the male. In other words, as far as behaviour is concerned, women would be more sensitive to androgens than men. Alternatively, androgens in the female may be of vestigial origin, their functional role having come to depend on their aromatization to oestrogens. There seems little doubt that the post-menopausal woman relies to a considerable extent on her androgens for her rather meagre supply of oestrogens. This intriguing puzzle remains to be solved.

But what of the role of oestrogens? Studies of ovariectomized and post-menopausal women should help us on this point. So far, adequate evidence has been lacking. We can infer that the normal sexual responsiveness of the vagina, particularly in terms of the transudate response, is dependent on oestrogen, and the 'dry vagina' of the post-menopausal women can be reliably treated with exogenous oestrogens. But we do not yet know whether these oestrogen deficiency states lead to reduced sexual 'appetite'. The claims that they do not have been based on inadequate evidence and we must await further studies. Progestogens also remain enigmatic. At one stage they were thought to have a negative effect on female and male sexual appetite, but now this is far from clear and any effect is highly likely to be dose-related in a complex and perhaps non-linear fashion.

As yet the role of hormones in the sexuality of women remains obscure. That they play a part in at least some women seems beyond dispute. Perhaps over the next few years, pieces of the jigsaw will begin to fall into place.

THE ESTABLISHMENT OF SEXUAL PREFERENCES

If we are to understand how our sexual preferences develop, and at what stage they become established, we must give some consideration to the timing of development of both our sexual responses and sexual appetite. Obviously, sexual preferences cannot be manifested in a sexual vacuum.

Sexual preferences are protean and many of the more unusual types are discussed in Richard Green's chapter in this book. But the most basic component is the sex of our preferred sexual partner – whether we develop hetero- or homosexual preferences. Whilst the majority of us become unequivocally heterosexual, when and how we do so remains a mystery. But let us at least consider some of the possibilities.

The sexual response system, at least its genital components, is probably intact from an early age, with the exception usually of orgasm and invariably of ejaculation. Male infants experience erection and commonly stimulate their own genitalia as if they gained pleasure from doing so. Female infants also behave auto-erotically and probably show at least rudimentary vascular changes in the genitalia. Some children appear to experience orgasm, though what proportion is not known. Kinsey relied to a large extent for his information on childhood sexual responses on the reports of adults who were or had been sexually involved with children. He reported the fascinating observation that at least some pre-pubertal boys are capable of multiple orgasms in a manner more typical of the adult female. This ability obviously declines once puberty and the onset of ejaculation commences, though the minimal interval between repeated orgasm and ejaculation in the male tends gradually to lengthen with advancing years, allowing for the vicissitudes of varying experiences and situations.

We know less about the orgasmic potential of the pre-pubertal girl, though, on the basis of Kinsey's evidence, it does not show the dramatic upsurge on reaching puberty that is seen in the male. The female's maximum orgasmic capacity may and often does take years to be realized.

But if there is commonly potential for sexual response during childhood, there is little doubt that the 'appetite' or at least the tendency to respond increases dramatically around the time of puberty, presumably because of hormonal changes. At what stage, therefore, do we develop our sexual preferences? Is the male infant masturbating with an erection expressing a heterosexual preference? Or is it more likely that this is simply a piece of self-gratifying behaviour which has no connotations, even in infantile terms, of a sexual partner?

It is a curious fact that even in the animal world we are far from certain when, how, or even if, sexual preferences are established. Frank Beach has recently criticized a tendency to make spurious comparisons between animal and human sexual behaviour. He points out that most sexual behaviour amongst animals demonstrates 'complementarity' rather than clearly defined sexual preferences. In other words, a typically female reaction, such as lordosis (hollowing of the back), will tend to occur in response to a typically male mounting activity, regardless of the genotypic sex of either participant. It is true that a typical female will show a much more developed and hormonally sensitized lordosis response to mounting than the typical male. Also a typical male is more likely to show mounting behaviour than is a typical female. So the majority of such interactions are between opposite sex pairs. But to equate the lordosis/mounting activity of two male animals with a human homosexual relationship in which both males are predominantly attracted to and prefer males as sexual partners is, as Beach points out, a specious argument.

Two lines of evidence to support the notion of an innate predisposition to heterosexuality and homosexuality have been the genetic and the hormonal. Are sexual preferences genetically determined? The most fruitful source of evidence in man are studies on twins. These show that monozygotic twins have a higher concordance rate for homosexuality than dizygotic pairs. The concordance, which varies from study to study, is nowhere near 100 per cent (it probably averages between 40 and 60 per cent). One possible explanation for these discrepancies is that monozygosity predisposes to homosexuality. In one study, Heston and Shields using the Maudsley Hospital twin register, found that 6.1 per cent of monozygotic twins and 7.2 per cent of dizygotic twins were homosexual, making this explanation unlikely. They also presented evidence that being a twin of either kind does not increase the likelihood of being homosexual. It is therefore reasonable to conclude that genetic factors do contribute. But that does not necessarily mean that sexual preferences *per se* are genetically determined. An alternative interpretation is that monozygosity increases the likelihood of two individuals reacting to environmental factors in a similar way.

Current interest in the possible hormonal determination of sexual preferences stems from our knowledge of sexual differentiation of the brain. We know that in most animals studied the exposure of the brain and particularly the hypothalamus to androgens during critical stages of fetal or perinatal development alters the neuronal structure of various parts of the limbic system and reduces or eliminates the basically female response characteristics of the brain. The characteristics are manifested in the positive feedback response and the cyclicity of the menstrual or oestrous cycle, as well as in the likelihood of 'female-type' (lordosis) sexual responses and other sexually dimorphic behaviour patterns. This effect of androgens, or the lack of them, will occur whatever the genotypic sex of the animal, suggesting that genetic influence on sexual differentiation of the brain is at least in part mediated by these hormonal effects.

Thus it has been postulated that sexual preferences, in particular of homo- and hetero-type, may be determined by the structural organization of the developing brain and that 'anomalies' of sexual preference (e.g. homosexuality) could reflect abnormal sexual differentiation. The only evidence that supports such a notion was reported by Gunter Dörner. He found that a group of homosexual men showed some evidence of 'positive feedback' in their gonadotrophin response to oestrogens, when compared with heterosexual controls. The reasoning behind this study is to be questioned. Although these general organizational effects of androgens on the brain are accepted for many animals, the picture in primates and man seems to be different. Although sexually dimorphic behaviour may be affected in this way, the 'positive feedback' characteristic of the hypothalamus of female primates and women appears to be immune to fetal or postnatal androgens. Robert Goy and others have found that female rhesus monkeys, androgenized either prenatally or immediately postnatally, may show anatomical and behavioural masculinization, but eventually menstruate and ovulate in spite of this. Anke Ehrhardt has shown that human females androgenized *in utero* as a result of the adrenogenital syndrome, do show tomboyish behaviour during childhood but they also may become fertile subsequently, though with a somewhat delayed menarche. In testicular feminization, the androgen insensitivity syndrome, where a genetic male develops anatomically and psychologically as a female, 'positive feedback' does not occur in spite of insensitivity of all tissues to androgens. Dörner's interesting findings, even if replicated, are therefore of very uncertain significance.

It is possible, though as yet unproven, that early hormonal factors may influence the development of gender identity (i.e. the psychological sense of masculinity or femininity) and alter the individual's capacity to respond to the continuing crucial environmental factors that follow. The gender identity in these terms does not necessarily indicate sexual preference. The male with heterosexual preferences may be relatively effeminate, whereas a homosexual male may be ultra-masculine in other respects. Nevertheless masculinity and femininity are likely to interact with other factors in the social learning process and there is probably a greater chance of an effeminate boy developing homosexual than heterosexual preferences.

Several studies have looked for endocrine differences between homosexual and heterosexual men simply in terms of their circulating sex steroids at an arbitrary point in time. Not surprisingly, considering the wide variety of situational variables that influence steroid levels and the more or less total failure to control for these, there has been no consistency in any of these results. We must conclude that so far no predictable hormonal picture has been identified.

It is difficult to deny the importance of learning in the shaping of our preferences. The real issue is the extent to which innate factors may influence learning and when the learning occurs. The male rat may learn to prefer a female because she is more likely to give an appropriate and a satisfactory response. If so, are sexual preferences only established after sexual experience? For the human subject we obviously have to allow for the effects of imaginary experience and it has been suggested that the 'conditioning' effect of masturbation, in endowing the masturbatory fantasy with erotic properties, plays an important part in this process. But is the learning simply a reinforcement of an inherent disposition, what Seligman and others have called 'preparedness to learn'? In other words do we assume that there is not equipotentiality of association between various stimuli and responses? And if there is such 'preparedness' for learning sexual preferences, is this innate or is it established by an earlier stage of learning? Psychoanalytic theorists are of the opinion that early 'object relationships' and their disturbances prepare the individual for later learning and development of sexual preferences. Evidence for this is difficult to find.

There is animal evidence, more relevant than most to the human being, from the developmental observations on monkeys by Harry Harlow and other primatologists. Monkeys reared in isolation, particularly when denied the opportunity for skin contact with either mother or a member of a peer group, show abnormal sexual development. The abnormalities are more obviously in the animals' methods of sexual interaction – clumsy mounting, etc. – rather than sexual preferences *per se*. But these social–sexual skill deficits could in turn influence the development of sexual preferences by modifying the types of sexual partner or experiences that prove to be rewarding.

With this animal model in mind, we can briefly consider some of the strands in the development of human sexuality and how these may interact to influence our sexual preferences.

As already mentioned, we have from an early stage a genital responsiveness and a source of sensual pleasure. This becomes further activated around the time of puberty. In addition, we develop our ability to form and maintain close inter-personal relationships. During childhood, these two developmental strands appear to be relatively detached from one another. But in adolescence perhaps the most crucial stage in our sexual development takes place as we incorporate our sensual responsiveness into dyadic relationships to create our first sexual relationships. Added to this is a potential for the conditioning of genital responses to a variety of 'erotic stimuli', which is perhaps most evident in boys. Glen Ramsey, in one of the early studies associated with Kinsey, found that pre-pubertal boys commonly experience erection to a variety of non-sexual but arousing situations (i.e. either pleasantly exciting or frightening). By the age of 12 to 13 such non-sexual responses are apparently much less common. This suggests a phase of relatively nonspecific genital responsiveness during which learning may play an important part, leading fairly quickly to the discrimination of more specifically sexual stimuli. Fundamental to this learning is the meaning attributed to these responses by the boys themselves. Probably for most boys, contacts with the peer group inform them of the sexual irrelevance of most of these responses so that the arousing situations do not become endowed with sexual significance. But in some cases, especially those isolated from their peer groups, these non-sexual stimuli may become 'sexualized'. Thus some of the more anomalous sexual preferences may become established. But one can also see how, in a comparable way, genital sexuality may fail to become incorporated into dyadic relationships, or more subtly, only into certain types of relationship. Thus potential partners may be excluded on the grounds of gender, leading to hetero- or homosexual preferences, or on other grounds, e.g. excluding sex from 'loving' relationships. Obviously, there is scope for a variety of influences to operate through learning on this incorporation of genital response into dyadic and otherwise affectional relationships.

The potential for conditioning of genital responses in the male has been demonstrated in adults by Jack Rachman and Ray Hodgson. Using a classical conditioning procedure, they were able to produce, at least temporarily, an 'experimental fetish' in a number of normal volunteers. This potential must give further scope for variability in our developments. Some 'conditioned stimuli' are going to be easier to incorporate into affectional relationships than others.

There may be an important sex difference in this respect. The penile erection of the male is a very tangible response, of which the young male is acutely aware and which requires some meaning attributed to it. For the prepubertal and adolescent girl, genital responses could occur without being recognized. Inappropriate meanings are hence less likely to be attributed to them and, perhaps, a girl's sexuality is much less genitally oriented during this crucial stage of establishing sexual preferences. If so, the confounding effect of anomalous genital conditioning will be less in evidence. This possibility is strengthened by the relative rarity of anomalous sexual preferences of the fetish type in females. Homosexuality is of course common amongst women, but the potential for bisexual preferences or changing sexual preference from hetero- to homo-, and vice versa, is probably greater in women than in men.

The question of the stability or fixity of our sexual preferences is an intriguing one. Could it be that there are sensitive periods during our development, when the capacity for establishing preferences (e.g. conditioning sexual responses to specific stimuli) is maximal, following which further changes or learning are unlikely? This remains to be demonstrated and more recently acquired evidence would suggest, if anything, a greater potential for change of sexual preferences during adulthood than has been previously thought possible. It is an interesting fact, nevertheless, that some individuals establish anomalous or 'normal' preferences in early puberty which remain fixed, whilst others experience a gradual evolution of their sexual preferences over a longer period, and pass through a variety of different stages. The explanation for such a difference remains uncertain.

In this chapter I have emphasized biological processes. But it will have become clear by now that, in almost every aspect, human sexuality presents us with a complex psychosomatic interaction. In order to tackle the many problems that stem from human sexuality, the nature of this interaction must be grasped. To do so presents a difficult intellectual and emotional challenge and there is a tendency for people or fashions to polarize either towards the social and psychological or towards the physiological. In the treatment of sexual problems, for example, the vogue for physical explanations and treatments that prevailed in the 1920s and 30s gave way to a predominantly, if not exclusively, psychological approach in recent years. Now the pendulum is probably beginning to swing back and we may hope that some middle ground will be occupied.

In the past few years, our scientific knowledge has increased at a formidable rate, and undoubtedly in the next few it will continue to do so. Unfortunately, this seems to lead inevitably to increased specialization, and yet human sexuality, like certain other vital issues, requires a multi-disciplinary approach for its proper understanding.

SUGGESTED FURTHER READING

Sexual Behavior in the Human Female. A. C. Kinsey, W. B. Pomeroy, C. E. Martin and P. H. Gebhard. (Especially part 3.) Philadelphia; W. B. Saunders (1948).
Atlas of Human Sexual Anatomy, 2nd edn. R. L. Dickinson. Baltimore; Williams & Wilkins (1940).
Human Sexual Response. W. Masters and V. Johnson. Boston; Little Brown (1966).
The psychophysiology of sexual dysfunction. J. Bancroft. In *Handbook of Biological Psychiatry*, vol. 3, *Brain Metabolism and Abnormal Behavior*. Ed. H. M. van Praag, M. H. Lader, O. J. Rafaelsen and E. T. Sachar. New York; Marcel Dekker, in press.

Suggested further reading

The social origins of sexual development. In *Sexual Conduct*. J. H. Gagnon and W. Simon. Chicago; Aldine.

The relationship between hormones and sexual behaviour in humans. J. Bancroft. In *Biological Determinants of Sexual Behaviour*. Ed. J. Hutchison. London; Wiley (1978).

Man and Woman, Boy and Girl: The Differentiation and Dimorphism of Gender Identity from Conception to Maturity. J. Money and A. Ehrhardt. Baltimore; Johns Hopkins University Press (1973).

Animal models for human sexuality. F. A. Beach. In *Sex, Hormones & Behaviour*. Ciba Symposium No. 63. Amsterdam; Excerpta Medica (1979).

3 Variant forms of human sexual behaviour
Richard Green

There is probably no limit to the number of 'variant' forms of human sexual lifestyles. People have great (frequently untapped) creative genius for evolving sexual scenarios. Many remain vague and are played out only in fantasy: others contain commonplace trappings and fall into recognizable patterns. But implicit in the assumption that we can define variant sexual behaviour is the belief that we can define the 'normal' or 'typical'. This false premiss beguiles many who would have law and order in their sexual universe; they define their own sexuality as typical.

The subject of this chapter lends itself rather easily to sensationalism, but I shall take care to avoid that here. Readers who look for erotica should consult the classic work *Psychopathia Sexualis* by Krafft-Ebing. Mine is a more sober view of things.

The topics I have selected for discussion are homosexuality, transsexualism, transvestism, sado-masochism, voyeurism, exhibitionism, rape, 'swinging' and group sexuality. Homosexuality was chosen because it is such a distinctive variant. Transsexualism and transvestism have received inordinate attention in the past, considering their relatively low incidence. Sado-masochism, voyeurism and exhibitionism are forms of behaviour that contain threads common to the sexual fabric of many of us. Rape (and paedophilia) are included because they worry people. Swinging and group sexuality worry people in a different way: most experience the positive pull of fantasy but consider the behaviour *too* daring.

HOMOSEXUALITY

People with an exclusive or major preference for a same-sex sexual partner make up a significant part of the population. Alfred Kinsey reported that four per cent of the men he

68

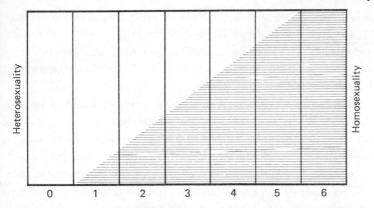

Fig. 3-1. Diagram of Kinsey's seven-point scale of sexual-partner preference, illustrating his concept of continuous variation between two extremes. 0 = entirely heterosexual; 1 = largely heterosexual with rare homosexual experience; 2 = largely heterosexual with frequent homosexual experience; 3 = equally heterosexual and homosexual in behaviour; 4 = largely homosexual with frequent heterosexual experience; 5 = largely homosexual with rare heterosexual experience; 6 = exclusively homosexual. (From A. C. Kinsey, W. B. Pomeroy and C. E. Martin. *Sexual Behavior in the Human Male*, Fig. 161; and A. C. Kinsey, W. B. Pomeroy, C. E. Martin and P. H. Gebhard. *Sexual Behavior in the Human Female*, Fig. 93. Philadelphia; W. B. Saunders (1948 and 1953).)

interviewed were exclusively homosexual throughout their lives, and another 13 per cent had been predominantly homosexual for at least a three-year period between the ages of 16 and 55. More than one male in three had experienced a sexual interaction with another male, after puberty, leading to orgasm. For girls and women, the rates were about a third of those for males. Estimates of the incidence of homosexuality in other countries are reasonably consistent with Kinsey's American figures.

Kinsey arranged sexual-partner preference along a seven-point scale, from exclusively heterosexual in fantasy and behaviour (0) to exclusively homosexual (6) (Fig. 3-1). He did not see heterosexuality and homosexuality as 'either-or' phenomena. 'Bisexuality' is a term with growing popularity and many meanings. In its strictest sense it describes people whose fantasies

69

and behaviour patterns place them at point 3 on the Kinsey scale. However, 'bisexuality' is also commonly used to describe individuals who have had sexual experiences or erotic fantasies with persons of *both* sexes, and they could be placed anywhere between 1 and 5 on the scale.

Homosexual behaviour appears to exist in most, if not all, societies. While attitudes vary considerably, in two-thirds of the societies described by Charles Ford and Frank Beach in 1952, same-sex erotic behaviour was considered normal and socially acceptable for at least some members of the community.

The mode of development of homosexual behaviour remains enigmatic (see also Chapter 2). The mechanism could include prenatal endocrine programming of the central nervous system, various genetic factors, features in the child's family life, such as resolution or non-resolution of the Oedipal conflict ('family romance'), or the impact of early socialization on 'social learning'. Sigmund Freud thought that everyone passed through a 'homo-erotic' phase in childhood, and then with resolution of the Oedipal conflict emerged as heterosexual; for Freud, homosexuality was an arrest of psychosexual development. Other authorities have rejected the idea of universal psychic bisexuality and argue that heterosexuality is the biologically normal developmental outcome in human beings. They maintain that traumatic childhood experiences lead to homosexuality – these experiences are typically (for the male) an over-intimate, possessive mother, and a passive, distant or rejecting father.

The psychoanalyst's view on female homosexuality typically includes lack of resolution of 'penis envy', together with unresolved Oedipal conflicts. In addition there could be a 'castration complex' and excessive early attachment to the mother, with considerable disappointment over the 'Oedipal wish' for the father, which is psychically compensated for by identifying with him and seeking mother substitutes. Alternatively, identifying with father may be linked with the selection of young women to serve as representatives of the self. Psycho-analytic theories about the development of homosexuality have been challenged on the grounds that they are based on clinical samples of homosexually-oriented persons experiencing conflict who seek out therapists, and because data from non-clinical samples generally do not support many of these findings.

A genetic theory of homosexuality is also held by some; it gets support from the work of Kallman who studied a group of primarily homosexual male twins, of whom half were fraternal and half identical. There was complete concordance in the sexual orientation of the identical twins (when one twin was homosexual, so was the other). By contrast, for the fraternal or non-genetically identical twins, the coincidence of homosexuality was not appreciably different from that of non-twin male siblings. Though there is some support for Kallman's findings, other studies have revealed monozygotic twins who were discordant in sexual orientation. Clearly we have not enough information to resolve the problem. Amongst other things, we need observations on twins separated at an early age and raised apart, and on children of homosexual parents brought up from earliest years in heterosexual homes. At present, it is difficult to be sure about the 'genetic loading' behind a homosexual orientation.

Neuroendocrine theories of homosexual development are currently getting close attention. This is because of the development of sophisticated laboratory techniques for measuring minute levels of hormones, greater understanding of the hypothalamic–pituitary neuroendocrine axis, and isolation of gonadotrophin-releasing hormones (see Book 7). However, reports of plasma levels of androgens and oestrogens in heterosexual as compared to homosexual males and females have been conflicting.

Proving more useful are studies on the hypothalamic–pituitary axis, specifically the effect on levels of luteinizing hormone of an intravenous injection of oestradiol. Men and women typically react differently and, interestingly, the response of homosexual males has been closer to that of heterosexual females. There is also some information on the effects of oestrogen priming on levels of luteinizing hormone and follicle stimulating hormone after administration of luteinizing-hormone-releasing hormone (LHRH). Female-to-male transsexuals (females who are sexually attracted to other females and want to be men) showed a response somewhat similar to that of men and different from that of women. Thus these early findings, though not yet confirmed, suggest that a prenatal endocrine state (e.g. elevated or depressed levels of androgen) may modify the fetal central nervous system so as to facilitate homosexual responses later.

We have already remarked that psychodynamic or early socialization theories have been advanced to account for the development of homosexuality. Recent attempts to confirm these findings have been conflicting. Some male subjects speak of a situation of a close, intimate mother and a passive, distant, hostile or absent father, but others do not. Of particular interest was Siegelman's finding that this typical early family constellation was more commonly found in both heterosexuals and homosexuals who scored high on neuroticism, and further that a higher degree of femininity (as measured by psychometric tests) in *both* homosexuals and heterosexuals influenced the degree of neuroticism. And so we see that a variety of factors other than sexual orientation *per se* may be contributory. Studies on the development of female homosexuality have also been inconsistent.

The behavioural features of male and female homosexuals are as varied as those for heterosexuals, and patterns of sexual experience differ from partnership to partnership. Some male and female homosexual couples live in a common household for decades, in what is for them a fully successful relationship, while others have only fleeting sexual contacts and are unable to achieve a desired long-lasting union.

More female–female dyads appear capable of existing over long periods with relative or absolute monogamy than male–male dyads. Fleeting relationships may exist in male homosexuality which are not found in the female homosexual relationship. These occur in the arenas of the steam bath and the public lavatory. The explanation for the greater stability of the female–

female relationship remains speculative, but the female in heterosexual pairs frequently represents the more monogamous partner, and furthermore less social stigma attaches to female–female pairs than to male–male pairs.

There is still considerable controversy over whether homosexuality in itself constitutes a mental disorder, and what kinds of (other?) psychopathology are to be found in the homosexually oriented. Recent scientific studies have led to acceptance of the idea that homosexuality *per se* does not constitute a mental disorder. In one investigation, no significant differences were found on any standard diagnosis of mental disorder for heterosexuals and homosexuals, except for somewhat more 'problem drinking' in the homosexual women. Another study led to the conclusion that 'Homosexual adults who have come to terms with their homosexuality . . . and can function effectively, sexually and socially, are not more distressed psychologically than are heterosexual men and women . . . and often those difficulties which are experienced by homosexuals are the result of social condemnation'.

This view reflects the thoughts of Sigmund Freud, who wrote that homosexuality 'is found in people who exhibit no other serious deviations from normal . . . whose efficiency is unimpaired, indeed distinguished by specially high intellectual development and ethical culture', and again: 'Homosexuality is assuredly no advantage, but it is nothing to be ashamed of, no vice, no degradation, it cannot be classified as an illness'. We know too that, despite rumours, homosexually oriented people are not significantly more likely to seduce children than heterosexuals. Nearly all the sexual abuse of children is by heterosexual males on female children, usually members of their own families (a phenomenon discussed in more detail later in this chapter).

Treatment

The great majority of homosexuals do not seek psychiatric treatment with the goal of achieving heterosexual orientation.

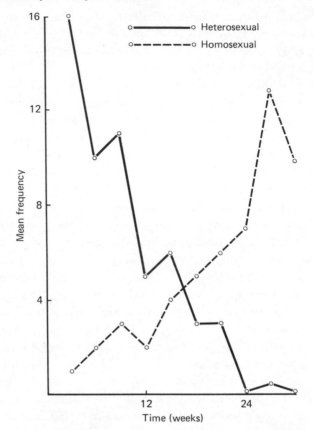

Fig. 3-2. Relation between homosexual and heterosexual impulses during the course of treatment. (From H. Leitenberg (ed.), *Handbook of Behavior Modification Therapy.* Englewood Cliffs, New Jersey; Prentice-Hall (1976).)

For the minority who do seek such orientation, several methods have been used ranging from psychoanalysis to aversion therapy. In one instance, after a minimum of 350 hours of therapy, about one-third of a group of male bisexuals or homosexuals became heterosexually oriented. In another, a course of twenty treatment sessions involving electrical shocks given simultaneously with homosexual visual stimuli increased the heterosexual arousal of more than half the subjects. Under treatment, homosexual and

74

heterosexual urges have been found in some studies to be inversely related (Fig. 3-2).

Another treatment stratagem which is gaining increasing acceptance is helping homosexuals adjust to their homosexual life-style.

TRANSSEXUALISM

Transsexualism has become the most widely sensationalized variant of sexual behaviour – possibly because of its spectacular nature and the fact that a number of transsexuals, including prominent people, have published their autobiographies. It involves more than a person merely adapting to the social and/or sexual role of someone of the other sex, and can include hormonal treatment and surgical modification of anatomical sex characteristics. Transsexualism is not new: the desire to live, or the act of living, as a person of the other sex is a human phenomenon that dates back through much of recorded history and spans a multitude of cultures. An emperor of Rome, Heliogabalus, is said to have offered half the Roman Empire to the physician who could equip him with female genitalia. James Frazer, in the *Golden Bough*, noted: 'There is a custom widespread among savages in accordance with which some men dress as women and act as women throughout their life'. At the beginning of the twentieth century, Westermarck describing the original settlers of North America noted 'In every part of the continent there seems to have been, since Ancient times, men dressing themselves in the clothes and performing the functions of women'.

In the first third of the twentieth century there were a few case reports in the European literature of people who had undergone sex-change surgery and one popular book, *Man Into Woman*, told the story of a Dutch painter who underwent a series of surgical procedures toward becoming a woman, and ultimately died during an operation. In the 1950s and 60s worldwide publicity was given to a former US soldier, George Jorgensen,

Variant forms of behaviour

TABLE 3–1. Brief history of a male-to-female sex reassignment (From Christine Jorgensen, *A Personal Autobiography*, New York; Ericksson (1967).)

Christened George Jorgensen. Earliest memories appropriate to those of young girl.

Age (years)	
11–12	Effeminate mannerisms commented on by family and friends.
17	Keenly aware of differences in emotions and behaviour from other males. Realization of love for male friend, not overtly expressed.
19	Drafted into US Army as a GI.
22	Consults doctors but no real help offered; reads literature on endocrinology.
23	Increasing anxiety over gender status, takes medical technician's course, learns about oestradiol, begins self-administration – slight gynaecomastia and emotional improvement.
24	Blood hormone assays instituted by doctor, results justify further treatment with oestrogens – continued improvement in sense of well-being.
25	Castration performed, progressive increase in feminity. Now happy and confident.
26	Adopts female sex for passport and changes first name. Surgical removal of penis. Active life as female singer and entertainer.
28	Final surgery – construction of 'vagina' and 'external female genitalia'.
33	Proposal of marriage by male friend; accepts, but engagement ended by mutual agreement.

who underwent sex-reassignment surgery in Denmark, took the name Christine Jorgensen, and became an international celebrity (Table 3-1). In England, the story of the 'sex change' of a former air hero from the Battle of Britain into Roberta Cowell also aroused considerable attention. More recently in England, the autobiography of James Morris described his transformation to Jan Morris, a distinguished authoress; as a male he had scaled Mount Everest. In the United States an eye surgeon, Richard Raskind, recently underwent sex-reassignment surgery and then sued successfully to be permitted to play professional tennis as a woman. This led to her being eliminated by Wimbledon Champion Virginia Wade in the first round at Forest Hills, New York, in 1977.

The sensational nature of these reports must be viewed against the rather tragic and deeply conflicting lives led by these people, at least up to the point of obtaining sex-reassignment surgery. From the earliest years, they recall feeling as though they were 'trapped in the wrong body'. If male, these children are socially stigmatized; if female, more privately troubled. Adolescence brings increased inner turmoil as the individual becomes aware of same-sex attractions, yet not feeling himself or herself to be homosexual. Typically, these people will not engage in overt sexual experiences with someone of the same anatomical sex until given hormonal treatment, or perhaps not until after surgical treatment. Those who do take part in sexual activity prior to sex-reassignment do not consider the experience homosexual but rather as heterosexual, because they feel themselves to be of the other sex. They are expressing 'psychical heterosexuality' and 'anatomical homosexuality'.

The steps toward sex-reassignment include 'social passing' as a person of the other sex, and day-to-day living in the aspired gender role. This is generally required by responsible medical centres for at least one, or two, years before surgery is approved. For both sexes, appropriate sex hormone treatment is given during this period. For males, there follows castration, amputation of the penis, construction of a vagina and electrical removal of facial hair. For the female, there follows breast removal, hysterectomy, ovariectomy and vaginectomy, and perhaps phalloplasty (construction of a penis).

The surgical construction of female-like genitalia and breasts in a male is a relatively easier procedure than construction of credible male genitalia in the female. Making a penis is fraught with difficulty and it is not possible surgically to produce the equivalent erectile capacity, via blood infusion, as in the normal male. However, 'erection' may be achieved by either insertion of a splint or through a hydraulic implant.

Psychotherapy has been unsuccessful in modifying cross-sex identity, but may be of help in screening out people with unrealistic expectations of what will follow a 'sex change' and in helping individuals through the tortuous process of sex-reassignment. However, most people who request a sex change are not willing to submit to therapy directed at changing their minds about surgery. They seek psychiatrists and psychologists who will sympathize with and accede to their requests for surgery.

Follow-up data on operated transsexuals are beginning to appear in the USA, England and Sweden. Generally speaking, surgical treatment clearly improved the social adjustment of patients, and the majority are happy with the results.

TRANSVESTISM

Transvestites are males who periodically dress in women's clothes, and often experience sexual arousal in association with the thought or act of cross-dressing.

A survey was made in America on 504 transvestites identified through the subscription list of a magazine published for heterosexual cross-dressers (Table 3-2). Both fetishistic cross-dressers (experiencing sexual arousal with cross-dressing) and non-fetishistic cross-dressers were studied. Respondents completed a questionnaire that yielded data on physical and socio-logical characteristics, childhood history, marital relations and dressing patterns, and on related behaviour patterns such as transsexualism and homosexuality.

Two-thirds were married and a few were separated, divorced, or widowed; most considered themselves heterosexual, and only a small minority bisexual or homosexual. However, many of the subjects who reported themselves as 'bisexual' also denied any homosexual experiences, thus raising the possibility that because of their cross-dressing they considered themselves 'bisexual'.

78

TABLE 3-2. Percentages of 504 transvestites with various attributes (USA)

Married at the time of study	67
Separated, divorced or widowed	14
Self-classed as	
heterosexual	89
bisexual	9
homosexual	1
Keen on athletic activities	81
Had undertaken psychiatric treatment	24
Felt as 'a woman trapped in a man's body'	12
Felt as 'a man with feminine side seeking expression'	69
Felt as 'a different personality when dressed as woman'	80
Felt as 'just myself dressed up'	20
Fetishistic cross-dresser	12
Treated as girl when a child	4
Earliest age of cross-dressing, less than 5 years	14
5–10 years	40
10–18 years	37
more than 18 years	8

(From V. Prince and P. M. Bentler. 'Survey of 504 cases of transvestism', *Psychological Report* **31**, 903 (1972).)

Typically, transvestites had participated on athletic teams in school or university. When asked whether they had ever consulted a psychiatrist regarding transvestism, a few replied that they had, having undertaken 'serious treatment' or gone for 'a few visits'; however, most had *never* seen a psychiatrist for their cross-dressing. They simply considered themselves to be men with a 'feminine side seeking expression', 'a different personality when dressed as a woman', or 'just myself dressed up'. Female hormones were rarely taken but many subjects indicated that they would *like* to take female hormones.

Most of these subjects came from families that were intact up to their eighteenth year, and when asked whether their fathers' provided a good 'masculine image', said 'yes'. Of those who said 'no' approximately half replied that this was because of their father's absence, not because of his drinking or cruelty.

Variant forms of behaviour

The dominant parent within their early household was more often said to be father. Nearly all the subjects has been brought up exclusively as a boy (no cross-gender raising), but a few indicated that they had been 'kept in curls till longer than other boys'. Thus, these individuals' responses did not support the idea that cross-gender role rearing underlay the later transvestism. Cross-dressing was first remembered by some prior to the age of five, but the modal age was around ten. Transvestite husbands generally informed their wives of their cross-dressing interests prior to marriage. The attitudes of the wives showed a wide range: some were accepting and co-operative, and others antagonistic; some permitted cross-dressing only in the home, and in their presence, while others did not want to see the husband cross-dressed. Most of the subjects reported having fathered one or more children.

Another question put to the subjects was whether they preferred to wear women's attire during sexual intercourse. A nightie was a popular garment, next came panties, padded bra and stockings. For those who wore stockings, high-heeled shoes were also popular. A few said they liked to wear a full women's costume during intercourse.

Most of the respondents seemed comfortable with their cross-dressing and even hoped to expand their activities or were trying to develop their feminine self more fully. The wish to cease cross-dressing was rarely expressed, and yet a common admission was that at one time or another they had destroyed or given away all their feminine clothing and sworn to discontinue the activity.

Treatment

The most successful form of treatment reported for fetishistic (sexually arousing) cross-dressing has been aversion therapy, involving chemically induced nausea or electric shock in association with cross-dressing. The treatment may also increase enjoyment of heterosexual intercourse.

Transvestism versus transsexualism

In Australia, biographical data and penile volume changes in groups of transvestites and transsexuals were compared. The transsexuals were younger, more likely to be unmarried, more likely to cross-dress completely and more likely to report homosexual interests, a feminine gender identity and of course a desire for sex-change surgery. The transvestites, in addition to being more heterosexually oriented, were also more often sexually aroused by cross-dressing; half of them reported they first experienced sexual arousal *when cross-dressed.* The frequency and intensity of the sexual arousal that accompanied cross-dressing typically diminished with age. Distinctly more of the transvestites had had heterosexual intercourse than the trans-sexuals, but for homosexual experience the proportion was reversed; most of the transvestites were exclusively heterosexual with a Kinsey rating of 0, while nearly all the transsexuals were judged to be Kinsey 5 or 6. Penile volume responses to a cine film of nude men and women were measured. The transvestites scored in the heterosexual range (penile erection to nude females) while those in the transsexual group excelled in the homosexual range (erection responses to nude males). The transvestites also tended to think about themselves being bound while cross-dressed or binding their partners or causing them pain.

Thus it seems that we should think of transsexualism and transvestism as separate syndromes, although there is overlap in some individuals. Transvestism is difficult to understand. While cross-dressing is a behaviour of some homosexual males (especially 'drag queens'), transsexuals (so they can appear to be the person they 'truly' are), professional female impersonators (containing a mixture of drag queens, transsexuals, and others), plus other males in mock chorus-line reviews, for no one else is it sexually arousing and associated with heterosexuality. And, it is the sole province of males. Why?

Variant forms of behaviour

TABLE 3-3. Percentages of 245 male sado-masochists (SM) with various attributes (West Germany)

Had practised SM during previous year –	
at least once a week	20
not at all	15
SM as exclusive sexual practice	16
SM as occasional sexual practice	16
SM essential to reach orgasm	15
Had experienced orgasm with SM	45
Used self-torture during masturbation	28
Exclusively heterosexual	30
Bisexual	31
Homosexual	38
Had sought treatment for SM	10
Had attempted suicide	9

(From A. Spengler. *Archives of Sexual Behavior* **6**, 441 (1977).)

SADO-MASOCHISM

Sadism is the pleasurable association of inflicting pain with sexual arousal and masochism that of experiencing pain with sexual arousal. They often co-exist in the same individual.

In a study of sado-masochistic behaviour on 245 males in West Germany – contacted as advertisers in newspapers soliciting others interested in sado-masochism or as members of sado-masochist clubs – heterosexuals, bisexuals and homosexuals were found to be equally represented (Table 3-3). *There were no female subjects.* The sexual interests of the men's wives were often incompatible with sado-masochism, and many of the wives were not even informed of the behaviour.

Heterosexual sado-masochists had less opportunity to make contact via advertisements than did homosexuals, and more often cited prostitution as a means for contact. By contrast, information from friends, clubs and parties were named more often by homosexually oriented persons. Most of the homosexual sado-masochistically oriented subjects had an acquaintance with like interests. The median frequency of sado-masochistic experiences was about five times a year. Half the subjects had a

82

relationship that persisted for over a year, but with some it lasted less than one week.

Most subjects studied said they would want to be sado-masochists even if they could 'decide freely about it', and there was little enthusiasm for seeing a doctor or psychiatrist because of their activity. However a small number had attempted suicide. Usually, active sadistic or passive masochistic interests were not the predominant or exclusive orientation, but rather an alternation between sadistic and masochistic roles. Sado-masochism as an exclusive sexual practice was uncommon and few subjects depended wholly on sado-masochistic activities to achieve orgasm. Most frequently, orgasm could be experienced without any sado-masochistic activity.

Elements preferred in sado-masochistic sexual scenarios were canes, whips, and bonds. Some practised anal manipulations and others utilized 'torture' apparatus. Leather and boots were often esteemed as parts of attire, jeans and uniforms being less popular. Rubber garments or women's clothing were used by a few. During masturbation, some reported self-bondage, self-beating, and torture of nipples with clamps.

The time when the subjects first became aware of sado-masochistic desires varied: most experienced them only after the age of 19, but some at the age of 30 or later. Subjects who were conscious of the phenomenon 'coming out' after 25 years of age were significantly fewer in the heterosexual group than the bisexual or homosexual groups.

Treatment

Behaviour therapy techniques have been found useful. For example, one investigator asked his patient to induce an erection with the aid of a sadistic fantasy and then to begin masturbating while concentrating on an erotic pin-up picture. When the erection waned, the patient was instructed to renew the sadistic fantasy. The patient was further instructed to focus on the pin-up erotic stimulus as orgasm approached. After several sessions the original pin-up picture was used as the 'recall stimulus' to back up other typical sexual stimuli with the sadistic fantasy kept in reserve. The patient's inclination to respond to sadistic fantasy was eliminated by using aversive imagery in conjunction with sadistic fantasies.

Another masochist was treated by 'shaping' his erotic fantasies – he was consistently able to achieve full erections to a fantasy of himself being beaten by a man dressed in a loin-cloth. Slight modifications of this fantasy, such as dressing the man fully or exposing his genitals resulted in less erection, while imagining himself being beaten by a woman yielded a further reduction in response. Treatment began with the individual fantasizing being beaten by a naked woman, and then the fantasy was systematically modified. The imaginary whip was made gradually shorter until it was replaced by a slap with the woman's hand. The slap then metamorphosed to manual sexual stimulation with the patient imagining himself bound. Ultimately, in fantasy, he was more dominant and initiated sexual intercourse. After eighteen treatment sessions the patient was responding with erection to conventional heterosexual fantasy.

In the use of aversive methods with masochists we might well ask whether a usually 'noxious' treatment would in fact represent a positive element for these patients. To explore this possibility a procedure has been employed in which the subject was partly able to avoid or induce electric shock in a treatment situation. The subject avoided shock *both* in the presence *and* absence of sexually arousing stimuli. The patient also showed some early favourable treatment response, but relapsed. Three sado-masochists, one sadist, and one masochist were also treated with electrical aversion therapy (pairing electric shock with undesired erotic stimuli), and were considered improved at a two-year follow-up. Though these results indicate that masochistic tendencies may be treated by aversion therapy, there is a case of a homosexual patient with masochistic tendencies in whom erections to homosexual fantasies increased in association with electrical shock.

Sado-masochistic variant life-styles have become more public in recent years. National organizations and regional clubs catering to such interests have evolved in the USA and other countries, and there is no shortage of appropriately oriented stories and photographic magazines in 'adult' book stores. While a degree of both giving and receiving pain is a common component of much 'typical' sexual behaviour and while sado-masochistic fantasies are common in collections of erotic fantasies, true sado-masochists rely more heavily on such a sexual scenario and go to greater 'pains' to choreograph the sexual ritual. Pain and pleasure, and aggression and sexuality, have been traditionally linked both in Freudian theory and the neurophysiological mapping of brain centres. Perhaps a comprehensive clinical study of sado-masochists will clarify the inseparability in these people of these seemingly polar bedfellows.

VOYEURISM

Voyeurs ('peepers', 'Peeping Toms') are adult males who, for sexual gratification, look into private homes or into areas or rooms reserved exclusively for girls or women in the hope of seeing them either nude or partially nude, the observation being without consent. These people belong in a mixed group which includes the socio-sexually underdeveloped, mental defectives, situational cases of 'peeping', and habitual drunkards. A small minority appears to have inadequate heterosexual lives, the remainder being a group whose prying is less habitual and who generally have adequate sexual lives. Voyeurs rarely spy on relatives or friends, but seek out strangers. They tend to be the only child or the youngest child in their family and commonly have few sisters. Many have lacked female friends during childhood. They engage in relatively little petting in adult life and have few friends in mid-teens. A somewhat small proportion of them marry and a few have extra-marital coitus.

We can epitomize the voyeurs' sexual makeup as of somewhat

stunted heterosexuality. The average voyeur has his first post-pubertal 'peeping' experience at about 15. More than incidental homosexual experience is not uncommon, but this does not seem to be an important feature.

EXHIBITIONISM

Exhibitionists deliberately expose their genitals to females in situations where exposure is inappropriate. Studies have revealed that, during childhood, these people did not socialize well with either boys or girls, although they did engage in sex play. Masturbation played an important role among the married men. Pre-marital petting began late and pre-marital coitus included a considerable amount of activity with prostitutes. Relatively few married, and those who did had infrequent marital coitus. In both extra-marital and post-marital sexual relationships they again relied heavily on prostitutes. Approximately half were repetitive exhibitionists. The exhibition derived from a compulsive urge generally triggered by emotional stress.

The age of onset of exhibitionism could be as early as ten but was more generally between 16 and 25. A large number suffered from impotence part of the time, especially the repetitive exhibitionists. The sexual availability of wives or other females did not seem sufficient to prevent this type of behaviour. Exposure was almost invariably to strangers and not to wives, friends or acquaintances, and commonly to young girls. The penis was generally erect and the act made out of doors and at a distance ranging from two to several yards. The great majority of exhibitionists did not resort to violence. Some had either attempted or seriously contemplated rape, but generally speaking exhibitionists appeared harmless and futile people.

RAPE

Perhaps as few as 10 per cent of rapes are reported, and yet the authorities are told of over 60000 forcible assaults each year in the United States. Rape involves both sexual and aggressive behaviour – even with rape committed following the prolonged absence of sexual partners, as in war, typical components are aggression and humiliation. Rape is not, however, purely aggressive as this might be satisfied by physically beating the victim. For some offenders the act is primarily sexual, for others primarily humiliating; for many, the sexual act itself appears to be less important than the ritual of the event which is often carefully planned, rather than impulsive. The forced sexual element may be so important that pleasure is lost if the victim meekly complies.

Lack of availability of other sexual outlets is not the principal concern for many rapists; many are married at the time of the offence, and most have been married at some time, indicating that they do have the capacity to establish socially acceptable sexual relationships.

One assessment of rapists and paedophiles ('child molesters') included the use of a 'hostility' rating scale. Rapists scored significantly higher than non-sex offenders or paedophiles. The rapists were further subdivided into four sub-groups, based on the amount of violence used in their offence. Although the more brutally violent offenders tended to score higher, differences were not statistically significant. In another study involving brutally violent rapists, non-violent rapists, and non-violent child molesters, all three groups scored above normal on the hostility scale, the brutally violent rapists highest.

Contributory influences in the background of rapists have been reported to involve parental seduction, typically by the mother, early prolonged sharing of the bed with a sibling or a parent, as well as incest experiences with older sisters, aunts, uncles, and cousins. On the other hand, many offenders claim to have been beaten frequently and severely by their mother, father, or both parents, or to have suffered severe physical punishment by their mother while the father remained passive, dependent, or weak. We can infer that when the mother is cruel and sadistic the future hostility of the son is likely to be directed against women because of the pain and humiliation he feels he has suffered. Rape may also be seen as a defence against feelings of masculine or sexual inadequacy, as when there is a history of a strong, dominating, over-protective mother, with a virtually absent father. The rape attack then becomes the ultimate in masculine behaviour (as viewed by the offender). On the other hand, neglect by both parents has also been implicated, and there are even some rapists who describe their childhoods as indistinguishable from 'normal'.

Sexual potency during rape may be lacking and some rapists need victim resistance to produce arousal. In one test a small group, whose erectile responses to audiotape descriptions were examined, did not become genitally aroused with stories of mutually enjoyable intercourse, but did when force was mentioned.

An investigation on the association between alcohol consumption and rape revealed that a great majority were drinking heavily at the time of their offence. Some drank to overcome timidity, others to obtain a kind of 'psychic muscle building' and a third group were primarily alcohol addicts, with rape merely one manifestation of their social disorganization. Rapists and their victims might or might not have met before, usually the latter, but in a few instances they were actually related.

Rape and pornography

There is a rather widespread belief that exposure to pornographic magazines, books or films plays an important part in shaping the mind and character of those responsible for sexual assaults. The author and his colleagues conducted a study in 1971 of sex offenders, including paedophiles and rapists, and compared these to non-sex offenders for the extent of their exposure to pornography. This was assessed during the year prior to commission of the sexual offence for offenders, or the year prior to interview for non-offenders. Acquaintance with explicit erotica

88

TABLE 3-4. Sexual offences recorded in Copenhagen. Porno-graphic magazines were legalized in 1967, but had become in-creasingly available during the preceding four or five years

	1959	1964	1969	Per cent decrease
Rape	32	20	27	16
Exhibitionism	249	225	104	58
Voyeurism	99	61	20	80
Paedophilia	51	18	19	63
Other offences against females	419	307	147	69

(From B. Kutchinsky. 'The effect of easy availability of pornography on the incidence of sex crimes: the Danish experience'. *Journal of Social Issues* **29**, 163 (1973).)

during adolescence was also taken into account. Both rapists and child molesters reported *less* exposure to erotica *both* in the year prior to the offence *and* during adolescence. Other investigators have reported that convicted rapists showed no significant difference from other people in their response to pornographic pictures, and that they purchased pictures of this kind less often than the non-sex offenders. In Denmark, the removal of pornography from the obscenity statutes in 1967 was associated with a decline in the number of rapes, attempted rapes and sexual offences against children (Table 3-4).

Treatment

There are few adequately controlled data on the long-term outcome of treatment for habitual rapists. Many offenders find their way into the penal system where little or no specific treatment is available, while others, classified as 'mentally disordered sex offenders', are given 'indeterminate' sentences in prison hospitals where some therapy is attempted.

In Germany and England, use of the anti-androgen cyproterone

Variant forms of behaviour

acetate appears to be effective in conjunction with psychotherapy, but in the United States the drug has not been authorized for use with sex offenders. Behavioural therapies included application of disagreeable stimuli (usually electrical stimulation through a wrist electrode) simultaneously with visual displays of people in various stages of resisting sexual advances (initially deemed attractive by the subject). Other approaches involve training in social skills in which persons deemed lacking in social graces or the confidence to perform enticing behaviour in a socially acceptable manner are taught the gentle art of seduction. Other methods involve attempts to persuade offenders to unburden themselves of their aggressive feelings and hostility towards women by verbal expression rather than in action.

Paedophilia and incest

The act of 'child molestation' (paedophilia) is of deep public concern. There is also considerable public misunderstanding about the people most likely to commit sexual offences against children. 'Dirty old men' lurking about school yards and parks offering car rides or sweets to young girls is an *idée fixe* of many. So is the notion that homosexually oriented males are significantly more likely to abuse male children. One does sometimes find older men who are unable to defer gratification of impulses until a socially more suitable situation can be arranged. This either reflects a personality defect in an otherwise normal individual or else a degree of organic mental impairment. Such offenders are caught in a conflict between morals and behaviour; in general, they are moralistic and conservative and maintain a sexual life normal for their socio-economic status. Many are extremely guilt-ridden as a result of their sexual activity with children. Often the offences require prior consumption of alcohol. Most of these offenders are not physically dangerous since they do not use force and seldom attempt coitus.

Inquiry into the background of paedophiles (Table 3-5) has revealed a frequency of pre-pubertal sex play with other children

90

TABLE 3-5. Percentages of paedophiles with various attributes (USA)

Offender: known to child and/or child's family	75
a member of the child's household	27
related to child by heredity or marriage	11
a stranger to the child	25
repeated acts over period of weeks to years	41
married at time of offence	31
drank heavily before offence	30
had had sexual contact with adult male(s) during childhood	18
selected only female victims	53
selected only male victims	29
exclusively heterosexual in adult relations	76
bisexual in adult relations	24
Average age of offender	35 years
Average age of victims	10 years

(From V. de Francis. *Protecting the child victim of sex crimes committed by adults.* Denver; American Humane Association (1969).)

similar to that for the typical male, and some had had sexual contact with adult males during their childhood. The researchers noted that while their subsequent adult offence was heterosexual 'the important thing is that the early experience may have impressed them with the realization that adult males do sexually approach children. Nevertheless, this pre-pubescent experience did not apparently result in an undue incidence in homosexuality in later life'. Two-thirds of the sample had married, and rather more than half of these had extra-marital coitus at least once. Nothing outstanding was found in the proportion of the heterosexual offenders against children who had homosexual activity after puberty.

A striking feature is that most offenders were known to the child and/or the child's family; many were members of the child's own household, being either a parent, step-parent or mother's paramour, and a few were related to the child by heredity or marriage, though not living in the child's household.

Variant forms of behaviour

Only one-quarter were strangers. They were generally of the same racial background. Most victims were around ten years of age, and nearly all were girls. The children were generally coerced by direct use of force or threat, lures such as money or gifts being rather uncommonly used. Some of the mothers had been child victims and some of the fathers were past offenders, chiefly against other family members.

Parents appeared very commonly to contribute to the circumstances of the sexual molestation by an act of omission or commission (or by actual perpetration of the offence). They failed to provide 'proper control' or exert 'adequate supervision' when away from home, even under circumstances where they had reason to suspect that the child was vulnerable to sexual abuse. Children escaped emotional damage when parents provided assurance and emotional security to offset the effects of the event, but most parents failed to do so and even contributed to the attention-charged climate.

Father–daughter incest is known worldwide (see also Chapter 5). Typical offenders appear to be those who have experienced poor adjustment with parents and even worse relations between father and mother; there is very frequently a history of divorce and separation plus financial trouble; often the entire home life is one of turmoil. Most of these offenders seem to have indulged in a considerable amount of pre-pubertal sex play, chiefly with girls, and to have been rather ineffectual, non-aggressive people who drank heavily, worked sporadically, and were pre-occupied with sexual matters. There was commonly also a high interest in extra-marital coitus, a high incidence of masturbation while married, and strong sexual responses when thinking of females. Typically the offender's wife was out of the home working, so that he was often at home with their children. All these things evidently pre-disposed to the incestuous relationship with the daughter.

Father–son incest is relatively rare. In a survey on about 1500 families that included a male child and a male parent, there were ten sons who were molested incestually by a biological father or a step-father. The fathers generally had a history of alcoholism, and four were known to be violent and physically abusive to their children. None was known to have homosexual relations with other than the immediate family or close relatives. The investigators judged that there was a high degree of complicity on the part of the mothers: in several instances they had known about the sexual activity for several years before community recognition or direct confrontation by the child forced them to take action. The sons expressed intense aversion for the fathers.

The past history of men guilty of sexual offences against children generally reveals that they are either individuals whose sexual partners since adolescence have been much younger than themselves, or for whom the assault against the child represented the appearance of sexual interest in children *after* a period of adult–adult sexual relationships. For offenders who were mainly oriented to young children, there were more male victims; for offenders who had previous adult relationships, there were more female victims.

Thus even when offenders are attracted to juvenile males this is not an extension of adult male homosexuality. What is commonly admitted by these adult offenders is that they are strongly sexually aversive to adult males. The immature boy is seen as attractive for his 'feminine features', with the absence of secondary male sex characteristics such as body hair and large muscles. Investigators of one large survey remarked that in over 12 years of clinical experience in working with child molesters they had yet to see *any* example of a regression from an adult homosexual orientation to a paedophilic homosexual orientation.

Treatment

Some practitioners have seen little evidence that behaviour modification techniques can work, but others are more optimistic. On the theory that the sex offender is socially inept, adult seduction techniques are taught. Attempted treatments have also

TABLE 3-6. Percentages of 20 000 respondents to a wide survey on 'swinging' (USA)

Had engaged in 'swinging' frequently	1
Had engaged in 'swinging' once or twice	3.5
Considered 'swinging' a future possibility	30
Had not participated and did not plan to	60

(From R. Athanasiou. 'A review of public attitudes on sexual issues'. In: *Contemporary Sexual Behavior*, Ed. J. Zubin and J. Money. Baltimore; Johns Hopkins University Press (1973).)

included conditioning stimuli (such as electric shocks) associated with pornographic pictures of underaged individuals. Few long-term data are available.

'SWINGING'

People talked about 'swinging' in the late 1950s as 'wife-swapping', a term that implied a value system where the wife was a chattel of the husband, exchangeable for other 'merchandise'. There then followed a more egalitarian term 'spouse swapping', and then 'co-marital sex'.

The number of 'swingers' is perhaps best estimated from the *Psychology Today* survey on 20 000 respondents (Table 3-6). The male partner generally initiates the idea that the couple engage in 'swinging', while the wife's initial reaction is unenthusiastic. If the couple do become swingers, there is usually a gestation period of weeks to months in which the possibility is discussed.

The childhood and adolescence of swingers in one study were different from those of the non-swingers. They were more apt to be 'only' children or to come from a small family. Male swingers perceived their parents as less happy and reported more parental divorce than controls. They had a more *laissez-faire* family life including a liberal attitude towards sex, with fewer parental rules. Both male and female swingers said they were less happy as adolescents. An interest in sex by swingers became manifest at an earlier age, i.e. during later childhood, and they also began sexual relations earlier. Their rate of pre-marital sex was higher.

Engagement in swinging activities did not appear seriously to depress the coital rate of these couples, and many more had intercourse with their spouse four or more times per week than among the non-swinging group. Most asserted that their marriage had become better as a result of swinging, at least as far as their sex life was concerned. Other surveys have indicated that for every long-term swinger there are three to four drop-outs. Thus the typical reaction is to give it up shortly after initiation, and the main reason for abandoning swinging appears to be marital discord.

GROUP MARRIAGE

Over half the group marriages observed in one large study survived less than one year! One investigator inferred that 'some people are simply overcome by inter-personal overload', and pointed out that careers that often require mobility may have to be sacrificed to remain in the group, and further that legal problems, including property transference and zoning ordinances, work against group stability. Most participants entered group marriage for companionship, sexual intimacy, personal growth, love, and the desire to create a good environment in which to raise children.

The biggest problems in group marriages are difficulties of communication, and conflicts between values and personalities. Next comes jealousy, notably for the time spent with other partners under the commitment. Sex *per se* is not as important an issue as jealousy and such things as the primacy of sex relationships, as many individuals in group marriages are almost totally free of dysfunctional jealous responses, not because they do not care for their initial partner, but because they feel secure and are as much concerned for their partner's pleasure as for their own.

What of the effects on children? From the results of a large

Variant forms of behaviour

study on children and child rearing in communes, the conclusion was drawn that 'the children of multiple parents do indeed enjoy some advantages over those in two-parent families'. Children are reported to be self-reliant and exceptionally competent inter-personally, to be well integrated into their families and to value their contribution to their families. They are happy and confident and above average in their sense of worthiness. There is an interesting parallel here with children born and raised in Israeli kibbutzim (see chapter 5).

Thus, swinging and group marriage, while receiving considerable publicity, are a favoured sexual pastime for only a small proportion of the community. The ideas are probably much more fantasied than practised. Whether they will increase in popularity is conjectural. They do have the potential for providing sexual experiences in an egalitarian, non-deceptive manner for those for whom monogamy is taxing. A major question concerns the means whereby some people are able to cope successfully with mutual non-monogamy. This aspect of human pair-bonding has not been well researched.

Some might argue over the exclusion of some patterns of behaviour from this chapter on variant sexual lifestyles – such as bachelorhood, spinsterhood, Don Juanism and asexuality. But there is a limit to the space available here, even if not so clear an end to our subject. Perhaps a lesson would be taught if all other variants could be included, for they would complete the continuum with 'normal' sexuality. Then, the essential meaninglessness of the terms 'abnormal', 'atypical' and even 'variant' as applied to sexual lifestyles might be all the more evident.

SUGGESTED FURTHER READING

The discrete syndromes of transvestism and transsexualism. N. Buhrich and N. McConaghy. *Archives of Sexual Behavior* 6, 483–95 (1977).
Experience with pornography. M. Goldstein, H. Kant, L. Judd, C. Rice and R. Green. *Archives of Sexual Behavior* 1, 1–15 (1971).

Suggested further reading

Post-surgical adjustment of twenty-five transsexuals (male-to-female) in the University of Minnesota study. D. Hastings and C. Markland. *Archives of Sexual Behavior* **7**, 327–36 (1968).

Swinging. A. Henshel. *American Journal of Sociology* **78**, 885–91 (1973).

Co-marital sex and the sexual freedom movement. J. Smith and L. Smith. *Journal of Sex Research* **6**, 131–42 (1970).

Deviant Sexual Behaviour. J. Bancroft. Oxford; Clarendon Press (1974).

Homosexualities. A. Bell and M. Weinberg. New York; Simon & Schuster (1978).

Group Marriage. L. Constantine and J. Constantine. New York; Macmillan (1973).

Patterns of Sexual Behavior. C. Ford and F. Beach. New York; Harper & Bros. (1952).

Sex Offenders. P. Gebhard, J. Gagnon, W. Pomeroy and C. Christenson. New York; Harper & Row (1965).

Transsexualism and Sex Reassignment. R. Green and J. Money (eds.). Baltimore; Johns Hopkins Press (1969),

Man Into Woman. N. Hoyer. New York: Dutton (1933).

Exploring Intimate Lifestyles. B. Murstein (ed.). New York; Springer (1978).

Sexual Identity Conflict in Children and Adults. R. Green. London; Gerald Duckworth (1974).

4 Patterns of sexual behaviour in contemporary society
Michael Schofield

NEW ATTITUDES TO SEX

In nearly all societies sexual behaviour has been subject to strong legal and moral restrictions. There were sensible historical reasons for this. Illegitimacy was often a personal tragedy for the mother and an economic disaster in poor village and rural communities. If the father of the child was not the husband, all sorts of problems of succession were created and the bastard child was not entitled to any of the family's land, a very serious drawback in a subsistence economy. If the mother was not married, this placed a heavy burden on the relatives who had to look after her and her child.

The only stable institution for providing care for children was the family. In most civilizations, therefore, it was essential to have harsh laws against sexual intercourse with anyone before marriage and no one except the husband after marriage. There were very severe punishments, including the death penalty in some societies, for those who broke these rules. In Western cultures this strict discipline was reinforced by making all human sexuality taboo. Its existence was kept secret from children and the subject was never mentioned in polite society. Silence and punishment created a fear of sex, which was thought to be beneficial to the community because it helped to restrain young people from indulging in any kind of sexual activity.

For nearly 2000 years conversation about sexual matters was discouraged in most Christian countries. Even the nude human body was thought to be distasteful and pornographic. All this is in the past; the most remarkable change in the last few years is the way we are so much more open and frank about sexual matters. To talk about sex is now acceptable, even fashionable; whereas not so long ago it was daring and naughty even to mention sex in respectable company. But much is just talk; when we come to action, there still seem to be all sorts of obstacles in the way. Listening to all the chatter, most of us are left with the impression that everyone else's sex life is far more exciting than our own.

The most important cause of this change in attitudes is the development of efficient methods of birth control. Contraceptives have been known and used for many hundreds of years, but doctors, until quite recently, have resisted all efforts to spread knowledge about birth control 'lest it give encouragement to the lewd' (John Marten, 1702). In 1868, the *British Medical Journal* objected to the idea 'of assigning to medical men the intimated function of teaching females how to indulge their passions and limit their families'. Even in 1978 the same Journal published a letter from 149 doctors including Sir John Peel, surgeon-gynaecologist to the Queen, former President of the British Medical Association, complaining about a document issued by the government (DHSS circular HC-78-12) which laid emphasis on the need to provide special family planning services for young people.

Coitus interruptus (also known as 'withdrawal' or 'being careful') was the most usual method in the past, but the condom also has a very long history. Malcolm Potts (in Chapter 2, Book 5) doubts the story that it was invented by a physician named Dr Condom at the Court of Charles II, but he has no doubt that it will still be popular when Charles III sits on the throne. In the early part of the twentieth century, spermicidal chemicals were used to good effect with the cap and the diaphragm, but the most significant change was in the early 1960s, when the pill and the coil (i.e. the intrauterine device), the two contraceptives with very low failure rates, began to be widely used.

Contraceptives can now be obtained in Britain through the National Health Service and in theory anyone who needs family planning advice or supplies can have them just for the asking.

Contemporary sexual behaviour

Despite this happy state of affairs, there are estimated to be 200000 unplanned pregnancies in the UK every year. An unplanned pregnancy need not, of course, be unwanted, but the official statistics are still very depressing.

In 1978, there were still over 53000 illegitimate births (19819 to girls under 20) and 102000 abortions (28003 to girls under 20) on women resident in England and Wales. In addition there were 35383 premaritally conceived births; over a third of teenage marriages take place when the bride is pregnant. Not all of these were shot-gun marriages, but it is reasonable to suppose that the impending arrival of the baby hastened the wedding day in many cases. Young people who are pushed into marriages before they are ready are more likely to end in divorce; six per cent of couples marrying in 1971 under the age of 20 had divorced by 1975 compared with two per cent of couples in the 20–24 age bracket. In fact 83 per cent of all British divorces involve women who married under the age of 24.

In a recent survey from the Office of Population Censuses and Surveys, it was reported that 55 per cent of 16–35-year-old single women had not used any method of contraception. Sometimes it seems as if everyone is getting contraceptive advice and supplies except those who most need it. There were 200 births to under 16s in 1951, but now the annual number of schoolgirls giving birth is about 1400, despite the fact that over six pregnancies in every ten occurring among schoolgirls are being terminated. An abortion is still difficult to obtain without delay in some regions of England and Wales, but the number of girls under 16 getting one rose from 1729 in 1970 to 3376 in 1979.

ADOLESCENT SEX

In a period when attitudes to sex are changing quite rapidly, it is the young who are most likely to suffer. Legal and moral decisions are made by the older generations who tend to have reactionary views on sex; they exaggerate the dangers and forget the pleasures. It has always been an agreeable task to tell young people how they ought to behave and older people usually expect adolescents to be bound by stricter standards of morality than they adhere to themselves. Until quite recently unmarried adolescents were not made welcome at family planning clinics and were more likely to receive a moral lecture than contraceptive advice if they went to their family doctor. Teenage girls who have committed no crimes are still taken into custodial care because they are in 'moral danger', which normally means they enjoy going to bed with men.

This is very unfair, but these prohibitions are justified as long as there is a gap between the age when a girl first has sexual intercourse and the time when she starts to use contraceptives regularly. Even now that contraceptives are freely available, there will still be a gap of a few weeks, unless it is going to be official policy to encourage a girl to use contraceptives before her first experience of sexual intercourse. But any attempt to prepare these girls before they have their first experience is likely to be misconstrued.

Although most people now accept that some kind of sex education in schools is necessary, the proportion of sexually active adolescent girls using contraceptives is still quite small. Among the 497 schoolgirls (i.e. under 16) who came to the British Pregnancy Advisory Service in 1976, 60 per cent said they had used no birth control method when conception was believed to have occurred, compared with 40 per cent for all single women attending the Advisory Service. In a national survey in the USA, half the sexually active adolescent girls reported that they had made no attempt to prevent pregnancy.

We must decide whether we are going to encourage young teenagers to use birth control, or try to prevent them from having sexual intercourse altogether, because it is an unfortunate evolutionary fact that women become fertile several years before what is, for both mother and child, the safest time for birth. Although the onset of fertility ranges from the age of ten to the mid-teens, childbirth for adolescents is more dangerous for themselves and their infants than it is for women over 20.

Postponing the age at first birth would reduce maternal and infant mortality and morbidity. The obstetrical complications of mothers in their teens include severe anaemia, third trimester bleeding, and prolonged and difficult labour. Infants of young mothers are more likely to die than those of older women and the low birth-weight of the children of adolescents increases the likelihood of congenital defects including epilepsy, blindness, deafness and mental handicaps. All these risks are increased when there are repeated teenage pregnancies.

Adolescent motherhood also restricts educational and career opportunities. An unmarried pregnant girl also has to face the social stigma that still comes with illegitimacy, or the pressures put on her and her boy friend to get married even if they are not really suited to each other. Adolescent parents find they are tied to the home before they are ready to settle down to the task of raising a family.

The number of unwanted pregnancies amongst teenagers although declining remains considerable. Of the 466 pregnant schoolgirls seen by the British Pregnancy Advisory Service, 17 had already had previous abortions and three of them had previously given birth to one child. Why do so many of them get pregnant despite the availability of effective contraception? The provision of a good family planning service is only part of the answer to this question. Education, motivation, moral attitudes and social pressures are also important influences.

THE PERMISSIVE SOCIETY

The 1960s heralded the dawn of the so-called permissive society. Those who were part of the swinging London scene hardly ever used the phrase; 'permissive' was simply a term of abuse used by people who felt that moral standards had slipped. The term is interesting because it implies that certain activities are being *permitted* which should be stopped. In the past when the Church had its own set of ecclesiastical laws which were enforced just as rigorously as the civil law, we accepted that our moral guardians should prescribe what we should read or do or think, and there was a 'moral police' to stop us from doing what was wrong. In more recent times the feeling has been growing that Acts of Parliament should not, and cannot, control the moral sides of our lives.

Permissive it may have been, but it was not a sexual revolution. Even today there are still signs that the moral earnestness of the nineteenth century is not quite dead. The authorities are beginning to realize that permissiveness is only rarely restrained by legal sanctions, but they continue to deplore such behaviour. Every five years or so there is a show trial about an allegedly obscene book in which after much expense and publicity a publisher is usually acquitted by the jury to the indignation of the judge. But censorship continues almost unabated in a quieter way by the Customs who seize masses of material which they have decided (without a trial) is pornographic, and by the police in raids on booksellers who are required to show (at considerable cost by employing expensive lawyers) why their stock should not be destroyed.

In general, society is getting more broadminded about what a person is permitted to imagine, but not about what he might do. Soft porn is more widely distributed and permitted in popular magazines and newspapers. But there is still the need to keep up the pretence of moral decorum and this makes it difficult for most people to do what nearly everyone is talking about doing.

Young unmarried adults in metropolitan areas seem to manage to find their way round the complex sexual scene. The older adults have more inhibitions and more commitments: they are informed almost every day that the possibilities of free sex exist, but when they set out to find it, somehow it is always elusive or expensive.

In my research into the sexual behaviour of young adults, 400 men and women all aged 25 were asked if they had any particular problems about sex. Over half (57 per cent) of the group said they had a sex problem and some of them mentioned more than one. Of all the many problems mentioned, the one that was most frequently reported was anxiety over their own sexual performance. Another worry was a waning interest in sex, while others said they were concerned about masturbation or feeling over-sexed.

The present-day emphasis on sex in the popular press, advertisements, films, drama, fiction and magazine articles including specialist publications devoted to sex, all seem to produce a dual reaction in some people. On the one hand they feel they are living in a society where every sort of sex activity is permitted, and on the other hand, and perhaps as a consequence, they begin to feel that sex has become too commercial and rather disreputable. Despite the high sales of sex manuals, the proliferation of advice columns in newspapers and journals, and the establishment of special sex therapy centres in many towns, the sad fact is that many people are sexually discontented. The two main hindrances to sexual satisfaction seem to be faulty sex education and unreal expectations.

FAULTY SEX EDUCATION

The idea of sex education now receives official blessing from the Department of Education and other authorities, but it seems to make very little impact on the pupils who receive it. Researches using large national samples seem to show that the effects of sex education can hardly be detected. The reason why it is so ineffective is that most courses are designed to suppress sexual behaviour, not to help people enjoy it.

Research results have shown that friends of the same age are a far more important source of information about sex than parents or teachers. From about the age of eight, children will start to give each other misinformation about sex. If sex education in schools is started too late, teachers have to break through a wall of resistance because the boys and girls think they know it all and will not listen. In several surveys the teenagers complain that they have not learnt anything during sex education that they did not know already, yet further questioning in these same surveys reveals large gaps in their knowledge.

A sensible course of sex education involves far more than merely knowing about the anatomy and physiology of the genital regions. That information is easy to acquire; there are over a hundred published books on sex education – all with diagrams of tubes and pipes, emphasizing the plumbing side of sex. (Rather less emphasis is placed on how the penis becomes erect and so some girls cannot imagine how the whole thing can possibly work.) These two-dimensional diagrams of the internal reproductive system are neither useful nor pretty. All those sperms running up canals leave young people with the impression that they should not start to have sex until they have donned a pair of rubber gloves. This kind of sex education produces the sort of boy who can name every part of the female genitalia, but does not know the first thing about how to treat a girl.

Local education authorities and school managers fear that sex education might encourage the pupils to take an interest in sex; if it is too explicit, it might put ideas into their heads. So it might, for that after all is what education is for.

One difficulty is that sex education courses in schools are outside the ordinary curriculum and do not lead to exam results or certificates. A few years ago a speaker at an important conference suggested pupils should have to pass an exam on sex. Inevitably this led to comments and jokes in the press and a leading article in the *Daily Mail* asked indignantly, 'What about the feelings of the poor boy who fails his "O" levels in Sex?' A better question would have been: 'What about the feelings of his girl friend?' The fact is that about nine out of ten boys and girls leaving school today would fail an exam about contraceptive methods, and even about how to make love.

As there is no examination board to set a syllabus, the contents of a sex education course is solely at the discretion of the head teacher who may think it is less troublesome to leave it for the parents to deal with the controversial parts. But few parents contribute anything towards their children's sex education. In my research it was found that 67 per cent of the boys and 29 per cent of the girls had never at any time had advice about sex from their parents. The difficulty for some parents is that they try to conceal from their children that they have any sex life themselves and so questions about the subject are likely to cause acute embarrassment. While the head teacher may leave it to the parents, they in turn think the school's duty is to supply the controversial information.

Inevitably there are serious gaps in the sexual knowledge of most young people. Sometimes a girl is not warned about menstruation before it happens to her; this may be because the parents simply cannot believe that their little girl is growing into a young woman. Boys are not warned about the autonomy of the penis – its embarrassing tendency to get hard and erect even when the boy's mind is not on sex, for example when riding on a bus; boys do not seem to speak about this even among themselves and some boys fear they must be sex maniacs because they have this uncontrollable penis.

What information a child does get from parents often reflects their own anxieties. The exchange of information among children is often very inaccurate. Ideas picked up from the media correspond to public attitudes which are not always the same as private behaviour. The schools have this unique opportunity to give a straightforward account of the many things that the pupils need to know about and which they would only stumble upon desultorily if they were left to find out for themselves. But the sad truth is that the schools have failed to grasp this opportunity. No one leaving school today really knows all he needs to know about sex.

UNREAL EXPECTATIONS

The second reason why people so often fail to get the best out of their sex lives is the wide difference between what happens and what they are hoping will happen. Many of these unreal expectations are the result of the faulty sex education just discussed, and myths and misconceptions created by the media, advertising and gossip. Much of the sex talk among young women is about unreal romantic aspirations; among young men, it is usually tales about sexual prowess or scurrilous jokes. All these fragments of inaccurate information build up into a fantasy image of sex which is not at all like the real thing.

A great deal of disappointment and misunderstanding arises from the rigid male and female roles that our society imposes upon men and women. Although we are beginning to break away from these roles, it is still true that if you are a man, your sex urge is assumed to be more imperious, more easily aroused by objects, and more specifically genital. But if you are a woman, it is believed that your sex urge is weaker, responds only to approaches from a lover, depends more upon the emotional relationship, and requires longer physical stimulation before it is fully aroused.

This is what we are all taught and this is what we come to expect. Anyone who does not fit this stereotype is treated as abnormal. Men who think love is more important than sex are thought to be sentimental. Women who like sex more than they like a long relationship are treated harshly; this is especially true of adolescent girls who are hounded by the police and social workers for behaving in a way that would be quite acceptable for a boy. Men who are not eager to go to bed with every pretty girl are suspected of being queer; women who enjoy an active sex life are called 'nymphomaniacs'.

The underlying assumption is that male sexuality is fundamentally different from female sexuality. Without doubt there are some good biological reasons for holding this old-established belief, but this does not mean that everyone's full range of sexual potentialities must be constrained for all time just because this is how most men and women have behaved in the past. Circumstances have changed and behaviour will also change to fit the new situation. For no matter what the basic biological structures may be, sexual behaviour is 'socially scripted' as contemporary sociologists would say; that is to say, the individual learns how to be sexual as he or she learns other kinds of social behaviour. Obviously there are physical differences, including differences in hormonal functions at different ages, between males and females; but the evidence for psychological differences (and so inevitable unchangeable differences in social behaviour) is very weak.

Anthropological evidence shows that our ideas about male and female sexuality are far from universal. The dominant female, more highly sexed than the male, appears in many different societies. There are dozens of examples of other communities which show that women are no more reticent than men when seeking sexual satisfaction. In many cultures adultery was allowed on special occasions such as religious ceremonies or festivals. Intercourse with priests met with social approval in many communities. The Roman Saturnalia were occasions for indiscriminate sexual intercourse for women as well as men. Modern festivals are said to exhibit some of the same tendencies if stories about the carnivals in Europe and South America are to be believed. Hospitality was another excuse for women to exercise their sexual rights; among the Eskimos it was a religious law that allowed the wife to go to bed with a visitor. In Siberia the women of Koryak used to pester tsarist officials, especially postmen, to sleep with them. Even in the history of the predominantly patriarchal Western Christian culture there have been periods (e.g. the Renaissance) when seductive females, influential in the arts and government, were thought to be more highly sexed than the men.

In present-day society people assume without question that it is 'natural' for men to be more excited by sexual fantasy and by pornography. Women are expected to be more concerned about love and starting a family. Men 'have sex', whereas women 'have sexual relationships'. The old idea that men need more sex than women does not stand up to examination. Women can achieve more orgasms, can go on for longer and can be just as highly sexed as men.

William Acton, the eminent doctor of the Victorian era, thought the idea that women might enjoy sex as much as men was 'a vile aspersion'. In his book *The Function and Disorders of the Reproductive Organs in Childhood, Youth and Adult Age, and Advanced Life, Considered in their Physiological, Social and Moral Relations*, he wrote: 'As a general rule, a modest woman seldom desires any sexual gratification for herself. She submits to her husband, but only to please him; and, but for the desires of maternity, would far rather be relieved from his attentions.'

Old traditions die hard and perhaps there are still blushing brides who are told by their mothers that something rather dreadful is going to happen on the honeymoon but she must learn to put up with it because it is part of marriage. It still seems to be true that men and women are both misinformed about each other's capacity for sexual stimulation. In an interesting study, Griffitt found that men frequently underestimated the level of women's sexual responses and women almost invariably over-estimated the extent of male arousal.

In communities with Christian traditions (and in many others), the reproductive role of sex has always been over-emphasized. Women were brought up to believe that motherhood should be the highest achievement and sole aim. Modern women quite rightly rebel at being regarded as merely producers of children. They demand the right to choose their own mode of sexual activity.

The success of the movement for the emancipation of women has produced another unreal expectation. Some women are now very concerned that they do not have an orgasm during sexual intercourse, although they would not have worried about this in the past. Worst still, some men now feel guilty if their wives do not have an orgasm and this situation is likely to end up with the women pretending to reach a climax by groaning and sighing, just to please their lovers. Some sex manuals have even suggested that the sex is a flop unless both have orgasms at the same moment. This is wishful thinking. Simultaneous orgasms are nice when they happen, but they are not a common experience and certainly not an essential part of sexual satisfaction. There is more to sex than orgasms.

Another unreal expectation which has been the cause of much distress is the assumption that lasting sexual attraction is an essential ingredient of a happy marriage. Everyone is in favour of romantic love; it is projected by the media with all possible persuasiveness. But it is seldom made clear that romantic love depends on feelings of desire based on sexual passion, not on the Christian idea of love which involves giving and sharing and commitment.

Most boys and girls are led to expect that one day there will be this blinding revelation that heralds the dawn of true love bliss. If it does not come along soon, the young person feels deprived; in the words of the song, 'You're Nobody Till Somebody Loves You'. When it does happen, the young are surprised to find that this intense sexual passion does not last for ever.

The romantic myth encourages the belief that the best reason for two people to get married is that they have 'fallen in love' meaning they find each other sexually attractive. If they are unlucky, they will find that they are not really suitable partners for life and the whole affair ends in tears. If they are lucky, youthful passions may cool and the sexual side of the marriage will become less important while love and respect for each other develops and grows.

These unreal expectations – the rigid sex roles, the belief that the male sex urge is so much stronger than the female's, the mad search for orgasms, the confusion between love and sexual passion – are the cause of many misunderstandings and much distress.

PREMARITAL SEXUAL INTERCOURSE

Premarital sex has been quite common in this and other societies for hundreds of years, but only in the last few years have the authorities admitted that in certain circumstances this may be acceptable.

This is a remarkable change in opinion, especially among youth workers, local government officials and others in authority who were implacably opposed to all sex before marriage until quite recently. In particular, church leaders tried to present a united front over the necessity of chastity prior to marriage. In 1966, the Archbishop of Canterbury issued a statement: 'We believe that the Christian Church should say plainly that sexual intercourse outside marriage is less than the best kind of loving and therefore wrong.' It was not just the more conservative element in the Church that rejected premarital sex. Similar opinions were held by men like Bishop Huddleston and Bishop Montefiore who were well known for their liberal views on other matters. Less than ten years ago the Assistant Secretary of the British Medical Association declared: 'As a doctor I can tell you that premarital intercourse is medically dangerous, morally degrading and nationally destructive.'

These views now look curiously old fashioned, but there is still the strong feeling, among older generations especially, that it is only permissible between two people who intend to get married. This is based on the unspoken assumption that, if anything goes wrong (i.e. if the girl becomes pregnant), they can always bring forward the date of the wedding.

When the parents know that two young people are sleeping together, pressure is put upon them to regularize the situation by getting engaged. The result is that many couples drift into marriage, even though they are not certain they will be happy together, because everyone expects it and the decision not to marry somehow turns their previously acceptable behaviour into a rather sordid affair.

The old question the teenage girl used to ask was: 'How far should I go?' Now the question is more likely to be: 'Does he really love me?' What she is really asking is: 'Does he intend to marry me?' The confusion between sexual attraction and marriage suitability will continue as long as people are encouraged to think that a licence for sex can only be obtained in exchange for a promise of marriage.

It would be to the advantage of everyone, the couples concerned and society as a whole, if teenagers married later, and this is more likely to happen if they know that they can get all the sex experience they need without having to make long-term commitments.

A girl may decide to sleep with a steady succession of boy friends until one eventually becomes her husband; although this would seem to be a sensible and logical preparation for marriage, such a procedure would arouse much criticism and hostility. Moralists still disapprove of promiscuity, casual sex, loveless coupling, sheer lust, the morals of the farmyard. The name-calling goes on, but there are signs that people are beginning to see that promiscuity is not necessarily all bad.

For one thing, promiscuity facilitates sexual learning. In ordinary circumstances the least effective way of learning a skill is to be told how to do it without the chance of putting this new information into practice. A much more effective method is to find out how to do it for yourself with help and instruction from an experienced teacher. Hence modern education puts the emphasis on the discovery method – learning by doing. Clearly any headmaster who advocates this method for sex education is asking for trouble.

Young people have a strong inclination to try themselves out sexually, as in many other ways. All the talk, all the jokes and all the warnings excite their curiosity and, understandably, they want to find out for themselves what it is all about. But we are forcing teenagers into premature obligations which curtail further experience and growth. Not all young people want to make such a firm commitment before they have their first experience of sexual intercourse. They should be encouraged to meet a large number of other people before being required to make such important decisions and give long-standing undertakings. There is no reason to suppose that the less one searches, the more likely one is to find the perfect marriage partner.

Now that efficient methods of birth control have lessened the chances of an unwanted pregnancy, young people do not see the reason why sex before marriage should be so restricted. It is noteworthy that the biggest change in attitude towards premarital sex occurred among student groups in the first place. In my earlier research (1965) I found that students were less likely to have premarital sex than working class adolescents, but now it is the other way round. This suggests that the reasons for the change were not only more freedom and less decorum, but also, in part at least, it was an intellectual decision; the arguments of the older people against premarital intercourse were not sufficiently convincing. Indeed the economic and social arguments evaporate when the couple take adequate contraceptive precautions; only the moral arguments remain. Before very long, many people will agree with Anatole France who once shocked his readers when he wrote that of all sexual aberrations chastity is the strangest.

ADULTERY

The opposition to premarital sex with more than one partner is still strong, but it seems to be abating. But a tolerant attitude to extramarital sex shows fewer signs of altering.

Of course adultery is not a new phenomenon, it is as old as the institution of marriage. Those troubadours, the precursors of pop, were not singing about a girl and boy falling in love, but were making a proposition to someone else's wife.

Adultery was condemned in nearly all societies for sound rational reasons. Sexual relations outside marriage were prohibited because this was the best guarantee that the husband was the father of the children. Anything that threatened the monogamous marriage was strictly forbidden. Adultery has been a crime in the legal history of most countries in Western civilization.

The opposition to adultery has always been particularly important for a woman. She is the one who will produce the bastard child when things go wrong. If some of the men occasionally went to prostitutes or loose women, that could be overlooked because it was not such a threat to the stability of the marriage. So gradually there developed this double standard and adultery became a male indulgence. No wonder it is the women who are against adultery in every opinion poll and attitude survey.

It is generally said that men want to be polygamous and women monogamous. Women as well as men believe that this is a biological fact of nature, but even a cursory knowledge of sexual physiology contradicts this idea. After a man becomes sexually excited and reaches orgasm, he is exhausted and cannot easily be aroused again for some time. But a woman can achieve orgasm and be ready to start making love again within half a minute. Therefore the needs of a man can be satisfied to exhaustion point by one woman. But, biologically speaking, one woman could continue for a long time at a high point of sexual excitement exhausting a series of men.

There is also another reason for doubting the belief that men are irked by fidelity but women are naturally suited to it. After all monogamy is an essential part of our social, religious and legal systems; these have been devised and administered by men. So it is the men who have invented the notion that they are free to seek sex elsewhere but women must remain faithful to one man. Obviously there is a measure of self-interest in this male attitude to fidelity, but it was not entirely selfish before the days of efficient contraception. Child-birth was dangerous until quite recently, so the men could protest that they were protecting their women-folk from additional dangers by disallowing adultery. Birth control has now eroded the one-sided male control of sexual relations.

Most people condemn infidelity because it results in the breakdown of many marriages. But cause and effect are difficult to separate, for it is this attitude to adultery that produces the bitterness which occasions the breakdown. It is the condemnation as much as the infidelity that causes the unhappiness, especially for those individuals who believe that divorce is the only respectable conclusion when one partner of the marriage has been unfaithful. This is to act on the principle that no bread is better than half a loaf. Divorce is often a very inappropriate solution for an affectionate couple who enjoy each other's company, even if there are sexual problems.

There are a few signs that a new attitude is emerging and some married couples now view adultery as a problem that can be coped with and not as a sign that the marriage has collapsed. In time it may be possible to achieve a new kind of sexual relationship within marriage which is free from possessiveness and jealousy. A more indulgent attitude to infidelity is still a long way off, but the advent of effective contraception and the sexual emancipation of the modern woman must be weakening the abhorrence of adultery.

TECHNICAL PROGRESS

Most forms of human behaviour follow logical patterns which depend on attitude changes which, in turn, depend on new information and technical progress. The most outstanding influence upon sexual attitudes has been the improvement in birth control methods.

There has been a massive increase in the use of contraceptives over the last 25 years, particularly since the introduction of the pill (see Chapter 2, Book 5). The new contraceptives are far more effective than those used before 1960. Consequently the failure rate for all birth control methods has declined in recent years. It is estimated that the number of contraceptive failures has fallen between the late 1950s and early 1970s from 15 to seven for every 100 women during the first year of marriage.

But no contraceptive method is 100 per cent effective. Indeed, one woman is supposed to have become pregnant with every known method of birth control. This may be a slight exaggeration, but it is a fact that *all* known birth control methods, except sterilization, have significant failure rates. Researches have shown that most married couples want a family of two or three children, and it is important to note that this wish cannot be fulfilled by using existing methods of birth control without abortion or abstinence during the average reproductive lifetime. Furthermore all known contraceptive methods have some disadvantages and the most effective ones (i.e. the pill and the coil) have undesirable side-effects ranging from morbidity to mortality. Since an abortion in the early weeks is known to be safer than completing the pregnancy, it follows that the use of a totally safe but less reliable contraceptive (e.g. the condom) backed up by early abortion is the family planning programme with the least hazard to life.

Abortion is particularly important as a second line of defence against an unwanted pregnancy for older women. Recent studies indicate that the risk of thrombo-embolic disease and death increases significantly with the age of pill-users, and many doctors now advise women over 35 to discontinue the use of oral contraceptives. Unfortunately, pregnancy among women coming towards the end of their fertile period is more complicated. The perinatal mortality rates among women aged over 40 are three times higher compared with women in their 20s.

In spite of religious and legal sanctions, abortion has been practised since time immemorial. Ways of terminating pregnancies are mentioned in early Chinese and ancient Egyptian writings. Since then all manner of materials have been tried including camels' saliva, the chopped hairs of a deer, a paste made from crushed ants, and an emulsion of goat dung and fermented vegetables. Despite this unpromising history, legal abortion today is as safe and effective as any surgical operation can be. It should be regarded, therefore, as a useful adjunct to birth control, not as a last resort in an emergency.

Not only is abortion safer, but it is also less likely to cause emotional upsets when it is carried out in the first trimester of pregnancy. In the United States only 11 per cent of pregnancies are terminated after the twelfth week, in Sweden seven per cent, and in Hungary, less than one per cent; in Britain 17 per cent of pregnancies are terminated after the twelfth week. The reason why the percentage is so high in Britain is because the 1967 Abortion Act requires legislative procedures that inevitably delay the decision to terminate a pregnancy. Any attempt to simplify the law is thwarted by the loud protests of the anti-abortion lobby.

A report in 1978 from the Office of Population Censuses and Surveys showed that attitudes to sterilization are also changing. At the rates now prevailing either the wife or the husband in well over a quarter of all married couples will be sterilized by the time the wife is 35. This study concludes that the role of sterilization is changing from a last resort for parents with already large families, to a preferred method of contraception for many comparatively young couples soon after the second or third child.

An Oxford University study of 17032 couples throughout the country shows that the husband has the operation in two out of three cases. Vasectomy (male sterilization) is now a simple and inexpensive operation with a very high success rate (about 99 per cent). The Roman Catholic Church still regards it as 'intrinsically evil' and 'absolutely prohibited', but there are over 10000 operations a year in Great Britain alone.

As more and more couples are choosing to be sterilized, we need to know more about its social and psychological consequences, since the operation is irreversible in most cases. Some men feel it is a blow to their pride and equate masculinity with the ability to make a woman pregnant. Some wives object to their husbands having a vasectomy because it makes them more available to other women. But many couples have discovered that this once-only procedure, taking less than 30 minutes and requiring only a local anaesthetic, saves the inconvenience of having to use contraceptives. The friends of married couples who have chosen sterilization sometimes ask: 'What if one of your children died?' But children should not be regarded as replaceable and the womb is not a production line.

The technical improvements will continue. For example, the Population Council in America has developed a contraceptive pellet, the size of a grain of rice; this can be implanted under the skin of a woman's forearm to release the contraceptive agent at a constant rate till the pellet disappears, and it should last for two or three years. All sorts of other chemical combinations are being tried out in the search for an oral contraceptive for men or a postcoital (i.e. 'morning after') pill. The medicine men are busy investigating other possibilities, from non-surgical sterilization to test-tube fertilization. Others have decided it would be simpler and safer to study folk medicine; one tribe in South America takes the fruit of a climbing vine which grows by the rivers of Guyana and mixes it with other herbs to make it into a delicious contraceptive cocktail which, they claim, keeps their women free from pregnancy for eight months.

FROM ATTITUDES TO ACTIVITIES

There have been some remarkable changes in attitude; changes in behaviour tend to be more gradual. The famous Kinsey reports showed that sexual behaviour varied only slightly over the years. My study of teenage behaviour showed that a third of the boys and a quarter of the girls under 19 had experienced sexual intercourse. Ten years later Christine Farrell reports a slight increase in the percentages. Whether it is 30, 40 or 50 per cent is not very significant, but it is important to realise that a large proportion of teenagers do have premarital sexual experience. (When later surveys report a figure as high as 80 per cent, perhaps we will start to worry about the remaining 20 per cent.)

People change their minds more easily than they change their ways. Anyone who has tried to change the behaviour of young criminals, alcoholics or gamblers knows that external pressures have only a limited effect on ingrained habits. Even to change a person's allegiance to a political party is quite difficult. We should not, therefore, be too surprised to find that the fashionable frankness about sex is not always reflected in basic changes in sexual behaviour.

In some areas the technical possibilities are held up by strong moral attitudes. When these attitudes change, the effect on our sexual activities will be felt sooner rather than later. For example, early abortion is now so simple that the operation can be carried out quickly and safely in an out-patient clinic. But there are still powerful forces that want to make it more difficult, not easier, to get an abortion.

The controversies that surround the 1967 Abortion Act (which are discussed on pp. 145–6 of Book 5) are contentious enough, but to raise the issue of the age of consent is like trying to conduct a debate in a nest of hornets. The choice of 16 for the age when a girl is legally permitted to have sexual intercourse is quite arbitrary; there is no real biological justification for any particular age. Less than a hundred years ago the age of consent was 13. And yet the age when an adolescent reaches puberty is much lower than it was when the decision was made, following a scandal about child prostitution in 1885, to raise the age of consent to 16. A hundred years ago half the girls were menstruating by the age of 15, but now half the girls are menstruating by the age of 12.

At present we tend to make it difficult for young people to get contraceptives, but if they could be persuaded to use contraceptives right from the very first time they have sexual intercourse, then the consequences of adolescent sex would become much less dangerous. Furthermore there is no doubt that many girls and boys under the age of 16 are willing and eager to have sexual experience.

These are the pressures in favour of lowering the age of consent, but the opposition to this change is especially strong. When the Criminal Law Revision Committee was given the task of reviewing the laws regarding sexual offences, many individuals and organizations proposed that the age of consent should be lowered.

The National Council for Civil Liberties proposed a series of ages of consent instead of one age below which a willing participant is deemed incapable of giving consent. They recommended that a person over 14 should be legally capable of consent and proposed an overlap of two years on either side of that age so that when the partners were over 12 but under 16, a consenting sexual act would not be an offence.

This mild suggestion produced vociferous reactions including headlines like SEX FOR TEENIES SHOCKER – *Daily Mirror*, SEXUAL SUBVERSION – *Daily Telegraph* and TOO MANY YOUNG PEOPLE IN CHARGE OF THE MORALS OF OUR YOUNG HAVE BEEN BRAINWASHED

Contemporary sexual behaviour

BY MINDLESS SOCIOLOGISTS – by Rhodes Boyson in the *Daily Mail*. One lady ('battling Beryl Knight, mother-of-three from Harold Wood' – *Romford Recorder*) gathered signatures for a petition to put a stop to this LOLITA CHARTER.

Everyone seems to hold strong opinions about the age of consent, but very similar arguments can be used to support the age of 13, 14, 15 or 16. It may seem inevitable to the lawmakers that one particular age is arbitrarily chosen, but this appears ridiculous to young people who are told sexual intercourse is illegal one day and legal the following day. The proposal that the present age of consent should be lowered meets with stiff opposition because some people fear that sexual activities bring into play emotions that are too difficult for young people to manage. Parents are often reluctant to admit that their children have sexual feelings. In reality, adolescents reach puberty earlier and are physically healthier than ever before; it is not surprising that young people have sex on their minds almost as much as their elders.

Not only is there a campaign to allow people to start sex earlier, but the old are now encouraged to go on longer than ever. As the individual gets older, irreversible changes occur in the ageing body. This is inevitable and the sensible person does not expect to be so active. It is not just a matter of unsupple muscles and creaking bones; hormonal changes in the body are also part of the ageing process.

One of the problems faced by women in the years after their menopause is the loss of oestrogen brought about as the ovaries reduce their production of hormones. Doctors have noticed that in many women past the age of sixty the lining of the vagina becomes very thin. Consequently some women find that penetration and the friction of intercourse becomes painful. Fortunately it is now possible to make up for any lack of oestrogen stimulation with adequate hormone replacement therapy. More and more women are now receiving treatment to make up for the loss of hormone production with the result that they can continue to enjoy an active sex life.

One of the points that Masters and Johnson, the famous sex

therapists, keep emphasizing in their discussion of the factors necessary for maintaining sexual capacity is the regularity of sexual performance. 'The most important factor in the maintenance of effective sexuality for the ageing male is the consistency of active sexual expression', they assert. This applies to women as well as men. Many women develop renewed interest in their husbands, and have described a 'second honeymoon' during the early fifties as a result of the ending of any fear of pregnancy. Women over fifty have usually resolved most of the problems connected with raising a family, and often there is a significant increase in their sexual urge.

Those who give up sexual activities for long periods may have difficulty starting again, but those who have an active sex life can often continue until well past the age of 70. Regular sex keeps the body in trim and so preserves vigour and self-respect in the old. According to Alex Comfort in *The Joy of Sex*, 'The things that stop you having sex with age are exactly those that stop you riding a bicycle (bad health, thinking it's silly, no bicycle).'

Ill health in old age need not be a killjoy. Even a bad heart or kidney trouble need not be a bar to an enjoyable sex life although it may be necessary to take things more easily than in the lusty days of youth. The exercise equivalent of sexual intercourse is similar to climbing briskly up a flight of stairs; not, as some seem to think, the equivalent of a five-mile run.

Perhaps the pace is slower, but this does not mean that the sexual life of the old need be less enjoyable. Orgasms may be less frequent, but this is not always an essential climax to an enjoyable session. Many people, young as well as old, find it more fun to journey than to arrive.

SEX IN A CRYSTAL BALL

Although we cannot say for sure how soon a change of attitudes will turn into a change of behaviour, we do know that one nearly always follows the other, so this gives us a glimpse of the future and a chance to predict the turn of sexual events.

We can reasonably anticipate that, in the not too distant future, premarital sexual intercourse will not be regarded as a commitment to marry; that sexual infidelity will not automatically signal the end of the marriage; that early abortion will be regarded as an adjunct to birth control; that sterilization will become much more conventional; that it will no longer be thought undignified for old age pensioners to have sexual needs; that adolescent sex will become safer and more acceptable; and that eventually the age of consent will be lowered.

These changes can be predicted because we can separate sex for pleasure from sex for reproduction, now that effective methods of birth control are available. Most religious authorities now accept that sexual intercourse is permissible even if the objective is not to produce a child. Not long ago, the Pope was still insisting that sex not specifically for procreation was sinful, but now a married couple may 'with joyful and grateful spirit reciprocally enrich each other' – the Catholic way of saying it's now alright if you happen to enjoy sex.

But the moralists have not yet woken up to the fact that once you separate fun-sex from biological sex, many other strongly held beliefs lose their support. If we agree that the human genitals are not designed solely for procreation, then all sorts of non-coital activities can be enjoyed; for example, it is no longer rational to maintain that homosexual activities are unnatural; nor is there any logical objection to masturbation. If sex is for pleasure, then there is no longer any reason why two people should be committed to each other for life before they have sex together; in fact, they needn't even be in love, though it's probably more fun if they are – at least for the time being. Furthermore it is no longer sensible to uphold the idealized version of women as being more interested in motherhood than sex, because girls want sex for enjoyment just as much as men. Those who use efficient contraceptives can, despite their upbringing, make a clearer distinction between love and sexual passion.

Love is the deep involvement of one person with another and sometimes leads to starting and bringing up a family, which should be taken very seriously indeed. But sex by itself is a different matter. It is basically playful – the innocent enjoyment of the body. As it is no longer necessary to think about sex within the context of marriage and child-rearing, we need not take it so seriously and we can enjoy our sexual pleasures more casually and lightheartedly.

SUGGESTED FURTHER READING

Family planning services : Changes and effects. M. Bone. London; Her Majesty's Stationery Office (1978).

How many children? A. Cartwright. London; Routledge & Kegan Paul (1976).

My mother said : a study of the way young people learn about sex and birth control. C. Farrell. London; Routledge & Kegan Paul (1978).

Response to erotica and the projection of response to erotica in the opposite sex. W. Griffitth. *Journal of Experimental Research in Personality* **6**, 330–8 (1973).

The sexual behaviour of young people. M. Schofield. Harmondsworth; Penguin (1968).

The sexual behaviour of young adults. M. Schofield. London: Allen Lane (1973).

Promiscuity. M. Schofield. London; Gollancz (1976).

The resolution of teenage first pregnancies. M. Zelnik and J. F. Kautner. *Family Planning Perspectives* **6**, 74–9 (1974).

Abortion : ten years on. The Birth Control Trust (1978).

Eleven million teenagers. Alan Guttmacher Institute, Planned Parenthood Federation of America (1977).

Schoolgirl pregnancies. British Pregnancy Advisory Service (1978).

Social commentary: fifteen to twenty-five. *Social Trends* No. 8 (1977).

5 Constraints on sexual behaviour
C. R. Austin

Surely the most impeccable and least objectionable form of sexual congress can be defined as that involving a heterosexual unrelated consenting married sexually mature couple of the same race, intent only on procreation. In contemporary Western society, variations on this theme are indeed common, but almost all social groups in the world abide by some of these specifications, and some insist on all of them. Each of the conditions must have some sort of social justification and presumably exists to curb some innate tendency in human nature, but to many people, especially the young, the proscription of so much of what comes naturally must seem a hangover of Victorian prudery or the legacy of an outmoded religious tradition. No doubt this is true of the idea that sex is only for procreation (shades of St Augustine) but a closer inspection of sexual *mores* and practices in man, and patterns in animals, reveals evidence that the constraints, like the activities they curtail, are essentially 'biological' rather than 'moral' in origin. They take the form of intuitive counter-reactions, and generally it is clear that definite advantages thus accrue to the race. A logical consequence then is that the constraints have become incorporated in traditional standards of conduct – and in religious imperatives – which is no more than consistent with C. P. Waddington's dictum that what is successful under Natural Selection can *ipse facto* be deemed ethically 'good'. In this chapter, I shall examine the manner in which sexual constraints are represented further back along the line, explicitly or implicitly, in the totems and taboos of primitive tribes and in antecedent influences in non-human primates. We might even pick up primordia in non-primate animals. This approach should reveal how limitation of sexual activity has actually come to possess adaptive value and accordingly to play an essential role under Natural Selection (or more specifically, Sexual Selection).

MARRIAGE

The most common constraint on sexuality is that it should occur only within marriage. The widespread practice of marriage (or other solemn agreement) is consistent with the human capacity for long-term pair bonding and is important for two main reasons: the very long period of human juvenile dependence, which is best supported on a stable and committed family foundation, and the highly developed human drive to acquire material wealth, the orderly sharing of which (and ultimately their disposal from one generation to the next) requires the existence of formalized relationships. Marriage caters for both of these needs; some people nevertheless reject it because they see evidence of capitalist motivation, but surely they thereby throw the baby out with the bath-water.

The two features of human marriage named above are not restricted to human relationships – social systems in animals can show them too. Though many mammalian young (such as those of rodents and rabbits) are ousted at weaning, several non-human primate (and some non-primate) societies retain the juveniles and immatures within a family group, which can become quite large. Groups of mountain gorillas, for instance, may contain a dozen or more individuals, ranging widely in age from the newborn to the fully mature (Table 5-1). In baboons there is a form of family–troop intermixture (Fig. 5-1). In elephant herds, too, related and unrelated animals of all ages keep company (Fig. 5-2), though the young bulls evidently leave the herd on reaching sexual maturity. Protection of the young in a lasting family association is clearly not exclusive to human society.

Neither is concern for material gain in itself a wholly human character, for land tenure and occupancy of caves, burrows, trees, etc., are contested by many animals and birds under the

Fig. 5-1. A troop of baboons (*Papio ursinus*) on the move, showing its composition of several families, each made up of individuals of different ages. The three highest-ranking males with their families occupy the middle of the group. (From K. R. L. Hall and I. DeVore. 'Baboon social behavior.' In *Primate Behavior*. Ed. I. DeVore. New York; Holt (1965).)

TABLE 5-1. Composition of three mountain gorilla groups as observed by George Schaller in 1959–60 (A, B and C) and by Dian Fossey in 1967 (a, b and c). (From D. Fossey. The Behaviour of the Mountain Gorilla. PhD thesis, University of Cambridge (1976).)

| | Groups | | | | | |
	A	a	B	b	C	c
Silverbacks[1]	1	2	1	1	4	2
Blackbacks	2	2	1	2	1	3
Adult females	6	4	9	5	10	5
Young adults	·	5	·	0	·	2
Juveniles	4	1	2	1	3	4
Infants	5	4	7	3	6	4
Totals	18	18	20	12	24	20

[1] Male gorillas become silverbacked at about 10 years of age.

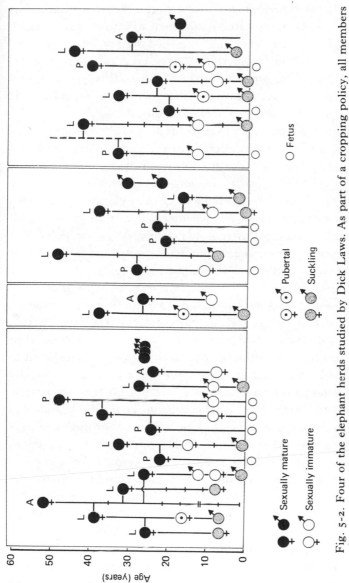

Fig. 5-2. Four of the elephant herds studied by Dick Laws. As part of a cropping policy, all members of a number of herds were killed and subjected to autopsy. A, anoestrous animals; L, lactating; P, pregnant. Placental scars signifying earlier pregnancies are shown by horizontal bars. (From R. M. Laws, I. S. C. Parker and R. C. B. Johnstone. *Elephants and their Habitats*, Fig. 7-10. Oxford; Clarendon Press (1975).)

name of territoriality (see Book 6, Chapter 3). Nest-building materials, too, represent wealth of sorts, and so do such courtship adjuncts as the remarkable assortment of objects collected by bower-birds – Jock Marshall recorded parrots' feathers and flowers, as well as fragments of glass, crockery, rags, rubber, paper, bus tickets and chocolate wrappers (Fig. 5-3). But of course there is no suggestion that these perquisites are inherited! In this practice human society is unique.

Despite the relatively rapid maturation of their young, long-term pairing is found in a few animals, such as the gibbon and indri among primates, and the kittiwake among birds (Book 6, Chapter 3). Brief 'consort' relationships are known in a number of species, involving only a male and female pair, but more often the group is larger. However, whether human marriage or animal relationship takes the form of monogamy, polygyny, polyandry or a mixed gathering, a common ingredient is the restriction of the right of sexual access to specific individuals.

The stipulation of 'married' as a condition for acceptable human sexual congress implies, in addition to positive aspects, disapproval of premarital and extramarital relations. In this area, reactions are most varied, a few human societies being unconcerned about either activity and many entirely tolerant to premarital intercourse. This, not surprisingly, applies especially where intercourse and pregnancy are not causally associated, which is true for a number of groups in the Pacific Islands and Australia. Some Australian aboriginal tribes, according to Ashley-Montagu, have the notion that when a woman becomes pregnant after marriage the child is derived as a spirit from a misty mythological past, and that intercourse merely prepares the woman to receive it. The child after birth is alloted a kinship linked with either father or mother, according to a traditional system (discussed later and exemplified in Table 5-3), and in addition becomes totemically grouped with individuals who are not blood relatives. Genetic connections are not appreciated. Tribes in contact with civilization realize, of course, the nature of stern reality, but often prefer to support the traditional

Fig. 5-3. A male satin bower-bird displays with his trinket collection before the female watching attentively from the bower. (Composite picture from several sources, especially A. J. Marshall's *Bower-birds: their Displays and Breeding Cycles, A Preliminary Statement.* Oxford; Clarendon Press. (1954).)

mythical view. Among the Trobriand Islanders, Bronislaw Malinowski found actual encouragement of premarital sex, in that common housing was provided for the young girls and bachelors, and from relationships initiated there more permanent commitments leading to marriage were supposed to develop, and generally did. Severe censure, however, was reserved for those responsible for childbirth out of wedlock, and no doubt too pregnancy was often the cue for marriage. Alternatively, in some Pacific tribes, such as the Areoi of Tahiti, the solution was infanticide.

Intercourse outside of established pairs or groups is often objected to in non-human primate species, the dominant males denying other males access to their mates, and the same kind of proprietory behaviour is seen among non-primates, such as elephant seals and topis (Book 6, Chapter 3).

Opposition in human society to extramarital intercourse, and to premarital intercourse with other than the future husband, could possibly be rooted in the belief that the intense emotional experience commonly involved forms a bond between the two individuals, and that this must detract from the strength of the ties so important for maintaining the association of marriage. (Certainly the growing popularity of pre- and extramarital sexual relations in contemporary Western society (see Chapter 4) is matched by a dramatic increase in divorce rates.) In ancient times too, great importance was attached to the emotional element in sexual intercourse. In *The Golden Bough*, James Frazer tells of a deep-rooted tradition in ancient Greece, whereby every woman being a virgin was bound by religious duty to sit in the temple of the Mother Goddess and offer herself for intercourse with a stranger; the practice was thought to be based on the belief that the deep emotional impact of this first sexual encounter – in the service of the goddess – would ensure a lasting devotion for the deity. It must have been something of a two-edged sword, for the first penetration can be painful, but the idea is interesting; moreover, it finds a parallel in the homosexual initiation rites for boys in some primitive modern societies.

CONSENT

Intercourse without free consent is rape and occurs commonly enough in human society (there were 1015 indictments in England and Wales in 1977, and it is evidently more frequent in the USA – see Chapter 3). In primitive communities, it seems to be associated mainly with inter-tribal warfare. Technically, rape could be said to occur in arranged marriages where the bride, often still a child, is under social and economic pressure. This would be so in England especially, for under the Sexual Offences Act of 1956 a girl under 16 cannot give consent in law.

Males are the physical superiors of females in many non-human primates and non-primates (Fig. 5-4), and would appear to be in a position to compel intercourse at will, but generally are not known to do so. Coitus by force is said to occur occasionally in orang-utans.

The uncommoness of rape in species other than the human is probably owing to the greater dependence of their sexual behaviour on hormone action (Book 2; Book 4, Chapter 2; Book 7); sometimes appropriate pheromonal and somatosensory stimuli are needed to arouse the male sufficiently for courtship. In human beings the cerebral cortex has taken over much of the control of sexual behaviour (though men depend on their androgen level to maintain libido and potency). This has disadvantages because cortical function is highly subject to conditioning by environmental influences and is wide open to the pressures of life in modern society. Human sexual behaviour is thus very much a product of the social environment and if this is sufficiently unfavourable excesses like rape ensue. In expiation, one could point out that man is the only species in which the female is in constant oestrus, and this must make a difference!

There is evidence too, that in many instances rape has a strong punitive (even sadistic) component (see Chapter 3) and represents a male protest against the escape of women from male domination, as epitomized in the progressive success of Women's Lib.

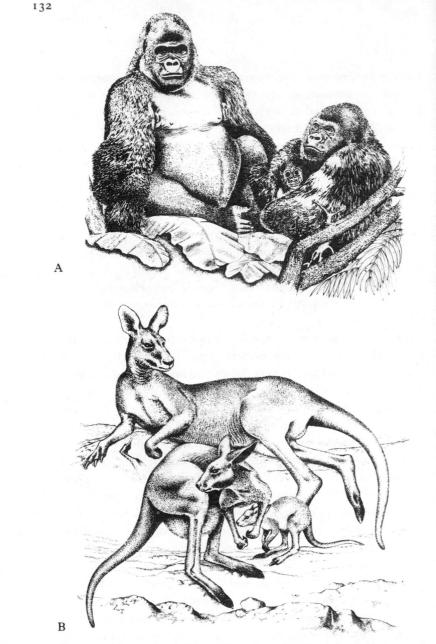

A

B

Fig. 5-4. Mountain gorilla and kangaroo families; the male parent is conspicuously larger than the female. (A, composite picture from several sources; B, from *Macropods of NSW*, wallchart issued by the NSW National Parks and Wildlife Service.)

Historically and ethnologically, male domination is the norm, as in polygynous non-human species. Female emancipation in our society is integral with our conception of human dignity and the rights of the individual; but it could, for all that, be in conflict with basic features of human nature. Arguably, if male domination is the product of Sexual Selection, it must be 'naturally right'. Departure from the principle reflects perhaps our capacity to 'evolve' intellectually in directions of our own choosing. There is a moral dilemma here; for this to be resolved sexual equality must prove to have survival value for the human race.

SEXUAL MATURITY

A preference for sexual maturity in couples liable to beget children has, as Michael Schofield points out in Chapter 4, some justification, because adolescent pairs are less adequately prepared mentally and physically for the rigours of raising a family, and pregnancy in adolescent girls involves greater risks of infant and maternal morbidity and mortality. But all of these problems can be avoided with effective contraception or adequate personal restraint, so perhaps what is really needed is a change in social attitudes.

According to Malinowski, Trobriand Island children from the earliest stages were freely allowed, even encouraged, to attempt to imitate their elders in sexual intercourse. He believed that such early experience prepared young people for a better balanced and more natural approach to the start of full sexual activity during puberty. A greater measure of responsibility was expected with the advent of sexual maturity, and pregnant 'unmarried' girls faced strong condemnation. Heterosexual (and homosexual) relations between adults and children were firmly discouraged.

Sexual maturity determines female receptivity and attractiveness and male responsiveness, in non-human animals much more positively than in man, but nevertheless sex play by the very young, similar to that indulged in by Trobriand Island children, has been described in chimpanzees by Jane van Lawick-Goodall. Interest in sex differences was shown by infant chimps almost before they could crawl, and continued unabated through the juvenile and adolescent periods. Adult oestrous females would freely participate (squatting co-operatively) in the sex forays of immature males.

Western society (and no doubt others too) has something to learn from these two groups of primates. The firmness with which we keep any mention of sex out of the lives of our children is matched only by the depth of the ignorance and apprehension with which they initially view the prospect of having relations of any kind with members of the opposite sex (see Chapter 4). A feature that man shares with other primates, and to some degree with other mammals, is the lack of a detailed inherent mating pattern, and therefore he must *learn by experience* how to react with a member of the opposite sex – aided preferably by percept, as with them, though most people have to manage on precept alone.

SIMILARITY OF RACE

Traditional preference in Western society is for similarity of race, as well as of colour, class, educational background, etc. This is true more for the choice of marriage partner than sexual partner, though it does apply there too in some measure. In primitive tribal communities, choice of marriage partner may be limited by custom to the same tribe or to specified neighbouring tribes with whom peaceful relations are cherished or sought. The Indian caste system generally requires marriage between people of the same caste; a man may sometimes marry a woman of lower caste, but penalties attach to marriage of a woman to a man of lower caste. The major world religions encourage selection of mates of the same faith. These are all instances of 'assortive mating'. The opposite inclination, 'disassortive mating', is practised as a guard against incest, and we will consider this in the next section.

Fig. 5-5. A group of rhesus monkeys behaving antagonistically towards a stranger introduced experimentally among them – an example of xenophobia. (From C. H. Southwick *et al.* (1974), Fig. 8; see Suggested Further Reading.)

Assortive mating may function more overtly in the negative form – the rejection of dissimilarity. Charles Southwick and his colleagues have observed strongly antagonistic reactions to strangers in established groups of free-ranging rhesus monkeys, a response they refer to as xenophobia (Fig. 5-5). The reaction was shown to juveniles and adults of either sex but not to infants, most of whom were soon adopted by adult females lacking infants of their own. The xenophobic response to older animals was sufficiently severe to cause the stranger to flee the district, and in one instance where the animal, a female, failed to leave she was killed. Southwick considers it logical that xenophobic behaviour should have been recorded in animals with highly developed territorial and/or group behaviour such as howler monkeys, gibbons, Chacma baboons, Syrian hamsters, white-footed mice, timber wolves, African lions and herring gulls, but not in those with an essentially open social structure such as chimpanzees*, red deer*, bandicoot rats and jackdaws, and birds forming large mixed migratory groups. He maintains that the same kind of dual distribution can be seen among invertebrates; thus, xenophobia has been demonstrated in social insects such as bees and ants but is clearly not a feature of communities of locusts, butterflies, gnats, mosquitoes and many beetles, which nevertheless readily aggregate.

Xenophobia may well have played a part as an isolating mechanism in speciation (a process discussed in Book 6, Chapter 4) and provides striking evidence of the significance that can attach to small details in determining an animal's reactions to other individuals.

RELATEDNESS

A remarkable polarization is seen among human social groups in their attitudes to sexual congress or marriage between closely related people. One extreme is represented in the practice that has appeared in several different civilizations – the inhabitants of the Marshall Islands and Hawaii, and the aristocracy of the ancient Irish, Egyptians and Incas – of regular and consistent marriages between brother and sister. According to Herbert Maisch, it was especially common in the seventeenth and eighteenth dynasties of the Ptolemies, and extended over a period of more than 300 years. Fascinatingly enough, Cleopatra was simultaneously her husband's niece and sister. The high nobility of the Incas are said to have been maintained by brother–sister matings for fourteen generations, without the appearance of any deformity or impairment. Other incestuous relations have of course existed also: there has, for example, been

---

* There is a difference of opinion here, for other authorities maintain that male chimpanzees will often defend a home range shared with females, and will kill intruders, and that red deer stags will attack intruders, in or out of rut.

Fig. 5-6. Consanguineous marriages are much more common in southern India (shaded) than in the north. (From L. D. Sanghvi. *Eugenics Quarterly* **13**, 291–301, Fig. 1 (1966).)

a tradition of permitted marriages between father and daughter and mother and son among the Eskimos and in various peoples of Melanesia and Micronesia.

In the modern world we find especially complex social pressures in India, where some systems forbid marriage within the same section of a caste and others require that the partners be no closer related to a common ancestor than by the fifth generation (through the mother) or seventh generation (through the father); by contrast, yet others give strong preference to close relatedness.

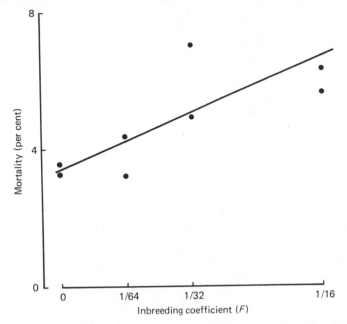

Fig. 5-7. Mortality rates in Japanese children correlated with their degree of inbreeding. (Second cousins have an inbreeding coefficient of 1/64, one and a half cousins of 1/32, the first cousins of 1/16.) (From W. F. Bodmer and L. L. Cavalli-Sforza. *Genetics, Evolution and Man*, Fig. 11.8. San Francisco; Freeman (1976).)

What has emerged through the mingling of many creeds and races is a broad division of the country into southern and northern regions where the consensus is, respectively, for and against consanguineous marriage (Fig. 5-6). A survey in Andhra Pradesh (one of the southern districts) revealed that two out of five marriages were between uncle and niece, or first cousins related through their fathers.

In many European societies, we encounter expressions of horror and disgust at the commission of incestuous acts, and in the old days penalties could be severe – enforced suicide or death by being thrown from cliffs or by burning. Incest was condemned by the Church as a 'monstrous and unnatural act'. There was irregularity, however, for a while under Pope Gregory III in the

TABLE 5–2. Observations on 18 children born of brother–sister or father–daughter matings, compared to 18 children born of unmarried mothers and unrelated fathers

> The mothers in the control group were matched as closely as possible with those in the incest group in respect to age, race, weight, stature, intelligence and socioeconomic status. (From M. S. Adams and J. V. Neel *Pediatrics* **40**, 55–62 (1967).)

	Incest group	Control group
Number in group	18	18
Average birth weight (kg)	3.01	3.19
Deaths	5	0
Major defects (evident by six months)	6	1
IQ	2 severely retarded, 3 about 70, 1 about 80	None < 80
Normal	7	17

eighth century marriage between people with anything up to the seventh degree of relationship (sixth cousins) was forbidden, by contrast, Pope Alexander VI in the fifteenth century blithely announced in a Papal Bull that he was the father of one of his daughter's children. Today, attitudes are still very varied, and some European countries lack legal sanctions altogether. In the UK the Sexual Offences Act of 1956 proclaims as an 'offence' sexual intercourse between a man and his granddaughter, daughter, sister or mother (and for a woman with her corresponding relatives), the penalty being imprisonment for seven years. Marriage between first cousins is not interdicted, which, under the circumstances, is surprising.

There are good biological reasons for discouraging marriage between close relatives, because of the greatly increased chances of homozygosity between recessive genes for deleterious characters. Mortality among children has been clearly correlated with the degree of cousin marriage (Fig. 5-7) and the same is true also for the incidence of certain anomalous conditions (Fig. 5–8).

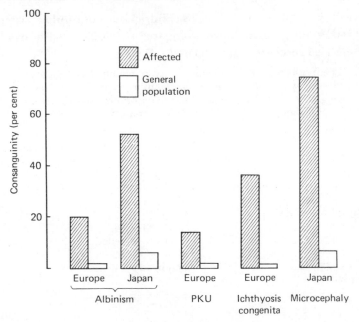

Fig. 5-8. The frequency of first-cousin marriages among parents of children showing certain recessive conditions ('Affected'), compared with the frequency among parents in the general population in different countries. PKU, phenylketonuria. (From W. F. Bodmer and L. L. Cavalli-Sforza. *Genetics, Evolution and Man*, Fig. 11.7. San Francisco; Freeman (1976).)

Even more dramatic effects were observed with brother–sister, father–daughter and mother–son matings (Tables 5-2 and 5-3). These facts point to the wisdom of avoiding consanguineous marriage, but we should note that, except for the results of brother–sister and parent–child matings, demonstration of the consequences of close-relative mating requires careful recording and rather sophisticated statistical treatment; remarkably enough, though, societies lacking these facilities have given evidence of such wisdom. Alternatively, their over-riding concern may have been to avoid sexual rivalry between close relatives, which was liable to disrupt family, and ultimately group, equanimity.

TABLE 5-3. Congenital anomalies in children of incestuous matings. Incestuous matings were father–daughter (88), brother–sister (72) and mother–son (1)

The control children were born to the same mothers as those in the incest group, but in this case the fathers were unrelated to the mothers. (From E. Seemanova. *Human Heredity* **21**, 108 (1971).)

Anomalies	Children of incest (n = 161)	Control children (n = 95)
Single malformations		
Hydrocephaly	1	1
Meningocele	2	—
Heart malformation (one with imbecility)	4	—
Cleft palate	1	—
Polydactyly	—	1
Talipes (with imbecility)	1	—
Luxation of the hip (one of incest group with imbecility, deaf mutism, retinitis pigmentosa and epilepsy)	5	1
Kyphoscoliosis (one with imbecility and epilepsy)	2	—
Dwarfism	1	—
Macro/microcystic kidneys	1	—
Hypospadias	2	—
Malformed external genitals	1	—
Cataract (with imbecility)	1	—
Multiple malformations		
Dwarfism, hydrocephaly, cataract (with imbecility)	1	—
Microcephaly, int. hydrocephaly	1	—
Microcephaly, luxation of the hip	2	—
Microcephaly, acetabular dysplasia (with retinitis pigmentosa)	1	—
Luxation of the hip, pyloric stenosis	1	—
Luxation of the hip, megacolon, hypospadias	1	—
Luxation of the hip, torticollis (with imbecility)	1	—
Luxation of the hip, ichthyosis and epilepsy (with imbecility)	1	—
Other abnormalities		
Imbecility, idiocy, uncomplicated	20	—

Anomalies	Children of incest ($n = 161$)	Control children ($n = 95$)
Idiocy, hydrocephaly	1	--
Imbecility or idiocy, epilepsy	2	—
Imbecility or idiocy, deaf-mutism	4	—
Imbecility, blindness	1	—
Deaf-mutism	2	1
Deaf-mutism, unilateral blindness	1	—
Epilepsy, uncomplicated	3	1
Congenital myotonia (with imbecility)	1	—
Haemorrhagic diathesis, unknown aetiology (with imbecility)	1	—
Adiposogenital syndrome (with imbecility)	1	—
Hyperaminoaciduria (homocystinuria, cystathionuria)	1	—
Mucopolysaccharidosis Sanfilippo	2	—
Totals	71	5

Particularly impressive are the kinship laws in certain primitive societies. James Frazer's classical analysis of exogamy among the Australian aboriginals is relevant here. He found that practically all Australian tribes practice exogamy, but that the details vary in complexity, some adopting a two-subclass division of the community, some a four-subclass, some an eight-subclass and some a sixteen-subclass. Marriages are required to take place between specified different subclasses (Table 5-4). Frazer believed that exogamy was taken up following a pre-existing state of promiscuity (for which, however, there is no reliable evidence) and because there developed a 'distaste for incest'. He pointed out that the two-subclass pattern avoided brother–sister marriages but not those between parents and children; the four-subclass pattern barred brother–sister and parent–child alliances but not marriage of first cousins; and the eight-subclass pattern excluded first cousins. Which all seems very commendable from a eugenicist's point of view, but whence came the desire to avoid incest among the Australian aboriginals who, as already

TABLE 5-4. Scheme to illustrate the principles involved in exogamy as practised by most Australian aboriginal tribes. In this instance, a system with four subclasses is shown. (Adapted from A. A. Abbie. *The Original Australians*, Fig. 19. New York; Elsevier (1969).)

TRIBE

MOIETY I		MOIETY II	
Subclass A	*Subclass B*	*Subclass C*	*Subclass D*

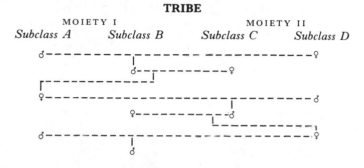

Patrilineal descent

A man in Subclass A can marry into C or D; if D, the child (say a son) goes to B (father's moiety). In turn, the son's choice of marriage partner must be only in C, and their child (say a daughter) goes to A, and so on.

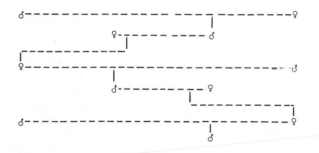

Matrilineal descent

With matrilineal descent, the principle is the same but the child goes to the class other than his mother's in his mother's moiety.

TABLE 5-5. The distribution of marriages in 125 couples of whom one or both were members of kibbutzim. 'Peer groups' are made up of individuals the same age, who were born and raised together. 'Movement' implies adherence to membership and ideology of the kibbutz system. (From Y. Talmon. *American Sociological Reviews* **29**, 491–508 (1964).)

	Percentage
Intra-second generation	
(*a*) Intra-peer group	0
(*b*) Inter-peer group	3
Intra-kibbutz	31
Inter-kibbutz	23
Intra-movement	27
Extra-movement	16

mentioned, traditionally did not recognize a blood relationship between parents and children? A satisfactory answer has long been sought, and may now be emerging.

Frazer's idea of a distaste for incest could mean that people might 'feel disinclined' to have intercourse with someone they had grown up with, and in more recent years observations on the occupants of Israeli kibbutzim have indicated that this could indeed be so. Yonina Talmon found that in a total of 125 marriages among second-generation inhabitants of three long-established kibbutzim, there was not one instance of marriage between members of the same peer group and only four marriages (three per cent) between members of different peer groups in the same kibbutz (Table 5-5). Kibbutz children are born and raised in similar-age groups as though members of a single household. They sleep, eat, play games and attend lessons in their groups, visiting their parents at home for only a limited period each day. According to Joseph Shepher, group solidarity is strongly encouraged, and individuality and competitiveness deprecated. Sex play is not interfered with and *is intense during early childhood*, but with entry into the teens there is a transient period

TABLE 5-6. The distribution of partners in the premarital sexual relations of twenty-two males and nineteen females born and raised in a kibbutz. (From J. Shepher (1971); see Suggested Further Reading.)

Partner	Males	Females
Same peer group	0	0
Other peer groups	1	1
Educated in the kibbutz	3	3
Adults of the kibbutz	2	7
Adults of other kibbutzim	4	4
From outside the kibbutz	12	4

of self-consciousness and embarrassment followed by a warm friendly relationship, lacking any notable sexual element. Shepher found no evidence of premarital sexual relations between members of the same peer group (Table 5-6), and, like Talmon, recorded no intra-peer-group marriages (he quoted census figures showing none in a large total of 2769 marriages).

Shepher believes that the findings are best explained on the basis of negative imprinting, pointing out that the following three important criteria are fulfilled: there is a definite and brief period of the individual's life when close coexistence can exert the effect (exceptions to the intra-peer-group bar are lacking only among kibbutz occupants who were there over the period between birth and six years of age), the effect is virtually irreversible (being evident even in those who subsequently married late or several times), and the effect is completed before the behaviour to which the imprinted pattern will be relevant is fully established (as witness the lack of premarital sex). These two sets of Israeli data are certainly thought-provoking.

Other reasons for incest avoidance are known. Jack Goody found distinctly diverse attitudes to incest adopted by two tribes in Ghana. For the Tallensi, sexual relations with a paternal aunt, sister or daughter were considered rather disreputable, but those with the wife of a man's own father, brother or son were severely punished. With the Ashanti, the judgements accorded to the acts were exactly reversed. The explanation, it seems, had to do with the disposal of possessions and the fact that the Tallensi are patrilineal while the Ashanti are matrilineal.

Jane van Lawick-Goodall recorded that sex activity between brother and sister tended to be avoided among her chimpanzees, and the same was true for the mother–son relationship. This was when all animals were sexually mature – by contrast, infantile and juvenile males practice copulation with their mothers. There is an instance reported of apparent refusal of mother–son intercourse in baboons, and Robert Hinde has inferred from published data in rhesus monkeys and Japanese macaques that although incestuous matings occurred, they were less frequent than would be expected by chance. To some extent this could be attributable to social behaviour – dominant males losing their position before their daughters are sexually mature, young sexually mature males freely moving between groups, and the apparent preference of females to mate with strange males. None of these devices is fully effective, however, and blood studies have revealed much more genetic diversity between than within monkey groups.

Among non-primates, brother–sister mating appears to be rare. John Maynard Smith considers that this is most commonly owing to dispersal of young males from the family group (which could, however, be contributed to by sibling reluctance to mate). Data from non-primates include those on deer mice which were found to reproduce sooner and oftener when paired with litter mates they had been separated from for a short time, than with those they had been with continuously, but soonest and oftenest when paired with unrelated animals (Fig. 5-9).

Some careful recent observations by Patrick Bateson on birds also have a bearing on this problem. Studying mating patterns in the Japanese quail, *Coturnix coturnix japonica*, he found that the males clearly preferred to mate with females of a slightly unfamiliar plumage (i.e. not siblings) than with those of familiar plumage (siblings). But they preferred both of these types to

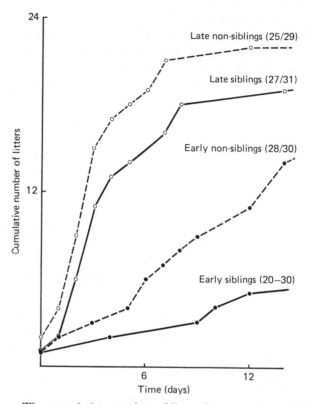

Fig. 5-9. The cumulative number of litters born to pairs of deer mice *Peromyscus maniculatus bairdi* after the birth of the first litter. The pairs of mice were either of siblings or non-siblings, each type being paired either early, namely at 21 days which is before sexual maturity, or late, at 50 days, after sexual maturity. The late pairs were kept as unisexual sibling pairs until placed with mates. (From J. L. Hill. *Science* **186**, 1042–4 (1974).)

females of a very unfamiliar plumage. Konrad Lorenz showed that birds undergo sexual imprinting soon after hatching and Bateson infers that this is expressed in such a way as to ensure an optimal balance between inbreeding and outbreeding – a complex pattern of behaviour that presumably has adaptive value.

As a phenomenon, resistance to inbreeding in animals, as in

plants, is reminiscent of the general preference for sexual over asexual reproduction, the virtue of sexual reproduction being that the recombination of genes distributed throughout a population augments the adaptive potential of a species. We would indeed expect an inherent tendency for exogamy rather than endogamy to prove successful under Natural Selection. At the same time there is obviously a counteracting force that works towards endogamy. There are biological advantages to be gained from preserving the integrity of successful sets of genes, and behaviour favouring this result would therefore also have survival value under the right circumstances. Nature, faced with such a dilemma, seems to waver between the two opposing goals.

HETEROSEXUALITY

The final specification for the 'ideal' sex relationship has to do with the notion that the participants should be a man and a woman, and generally speaking this is so. However, something like 10 per cent of the UK population are avowed homosexuals and the actual proportion could well be higher than that. This would include many who are bisexual in behaviour. In any event, homosexuality has been known as long as prostitution and is certainly a feature of contemporary society in all countries. Homosexual inclinations could be an important component of social stability, contributing materially to tribal solidarity in primitive groups and to *esprit de corps* in many Western institutions: one-sex schools (both male and female), social clubs, units of the armed forces, sports groups, etc.

Homosexual activities are known to occur in natural, non-human primate communities, especially between males. Generally speaking these appear to be uncommon and incomplete acts, but as Art Riopelle points out, homosexuality could very well be the norm in the bachelor groups in which the young males live until they can begin to form a harem. Terry Maple has also described the occurrence of mounting between females, mutual genital stimulation and oral–genital stimulation between male

148

Fig. 5-10. Mutual genital stimulation and unilateral oral–genital stimulation between male stump-tailed macaques *Macaca arctoides*. (From T. Maple (1977); see Suggested Further Reading.)

and female, and between males (Fig. 5-10), and he records that these sorts of activities have been seen in chimpanzees, gorillas, rhesus monkeys, spider monkeys, Japanese macaques, vervets, pigtails and tree shrews. Though seen in the wild, they are much more likely to be expressed in captivity. Among non-primates, homosexual behaviour has been described (it is common in cows, as every farmer knows) but is generally less evident than in primates. As we have had occasion to remark before, in animals below the higher primates, female (and to some extent male) sexual behaviour is more subject to endocrine control and dependent sex-specific stimuli. Another point is one made by Frank Beach, namely that man is the only species in which the sexual responses of male and female are almost identical – sexual flush, orgasm, pelvic thrusting, posture (male or female may be dorsal), and so on. No wonder gratification can be got from homosexual contacts to a degree denied to non-humans!

Human sexuality shows much greater diversity in expression than that of any other animal, though most features are represented in some measure in non-human primates and to a lesser degree in non-primates. Man's partial escape from the hormonal control of his sexual behaviour, and its substitution in large measure by cortical control, has greatly increased the range of his expression and experience, but the switch has not been an unmixed blessing. Weighty considerations besides sexual gratification command his attention and these relate to the care of progeny, the avoidance of inbreeding, the inheritance of material wealth, and recognition of the complex social conventions governing the choice of sexual partner. In the main, these items represent responsibilities to society. Non-humans do not need to worry about such things; man alone has the onerous task of being master of his fate and captain of his soul.

SUGGESTED FURTHER READING

Unusual sexual behaviour of non-human primates. T. Maple. In *Handbook of Sexology*. Ed. J. Money and H. Musaph. Amsterdam, New York & London; Excerpta Medica (1977).

Mate selection among second generation kibbutz adolescents and adults: incest avoidance and negative imprinting. J. Shepher. *Archives of Sexual Behavior* 1, 293–307 (1971).

Xenophobia among free-ranging rhesus groups in India. C. H. Southwick, M. F. Siddiqui, M. Y. Farooqui and B. C. Pal. In *Primate Aggression, Territoriality, and Xenophobia*. Ed. R. L. Holloway. New York & London; Academic Press (1974).

The behaviour of free-living chimpanzees in the Gombe Stream Reserve. J. van Lawick-Goodall. *Animal Behaviour Monographs*, vol. 1, p. 165. Ed. J. M. Cullen and C. G. Beer. London; Baillière, Tindall & Cassell (1968).

Bisexuality and the problem of its social acceptance. C. R. Austin. *Journal of Medical Ethics* 4, 132–7 (1978).

Coming into Being among the Australian Aborigines. M. F. Ashley-Montagu. London; Routledge (1937).

Totemism and Exogamy, vols. 1–4. J. G. Frazer. London; Dawsons of Pall Mall (1968).

Comparative Studies in Kinship. J. Goody. London; Routledge & Kegan Paul (1969).

Incest. H. Maisch. London: André Deutsch (1973).

The Sexual Life of Savages. B. Malinowski. London; Routledge (1929).

The Mountain Gorilla. G. B. Schaller. Chicago & London; University of Chicago Press (1963).

The Evolution of Sex. J. Maynard Smith. Cambridge University Press (1978).

6 A perennial morality
G. R. Dunstan

A morality is a means to the right use of the given. The given in a sexual morality are the nature of human personality and the nature of human society. Given also is the biological basis of human personality and of human society. It follows that a series of volumes on the biology of reproduction, including human reproduction, would not be complete without a chapter on morality. It follows also that the chapter on morality could not be written without regard to the fundamental biology.

To give biology this fundamental position is not to accord it a finally determining position. Biology shapes the stock on which the human personality blooms. It provides the physical capacities, elementary and advanced, for all those activities and responses that we call life. But the activities and responses that we designate peculiarly human – and there are some – transcend these capacities while employing them. There remain areas of human freedom not determined by biological impulse. Freedom permits choice, and demands it; and choice both presupposes and requires a morality, criteria for choosing. The point of departure from the biological exposition for the chapter on morality is in Austin's introduction of the cerebral cortex: 'In human beings the cerebral cortex has taken over much of the control of sexual behaviour' (p. 131). It is from this, with its openness to environmental influence, that flows the sublimity as well as the abysmal misery of sexuality in human experience. A sexual morality operates precisely to shape these environmental influences for what is conceived to be human good, human flourishing.

To claim something distinctive in human kind is not to disown what it has in common with animal kind, back through the highly complex biology to the basic biochemistry. This is part of the 'given' with which morality has to do. Contemporary sexual morality affirms it, consents to it, does not deny or repudiate it. Equally the claim that there is a specifically human manifestation of sexuality is not invalidated by anthropological studies that show considerable varieties of sexual relationship in different peoples at different times, or by studies describing the many varieties of a sexual bent that can occur in individuals. It is significant that Richard Green in Chapter 3 adds a section on 'treatment' to several of the 'variant forms' of human sexual behaviour that he describes. Part of the 'given' is a range of human experience of which there is ample evidence in literature, poetry, drama, music, art, in which sexual relationship, while fulfilling its basic biological purposes, extends beyond the genital into the erotic, the romantic, the intellectual, and the spiritual, so creating a culture and immeasurably enriching the common possession of mankind. Within this culture stands a morality, evolving but perennial. Sherwin Bailey has traced this morality in *The Man–Woman Relationship in Christian Thought*.

Sexual morality as a fact is perennial: sexual access is too vital a factor to leave unregulated in any community. Its shape and its prescriptions evolve: the evolution accompanies (though always a little behind) the evolution of the relevant knowledge or science, and it produces conceptual mutations of its own from time to time, as in the recognition of mutuality and consent. The morality closely concerns the basic biological purposes of sex as they have been described in the previous chapters. The most basic is reproduction; this is universal. With it, in some species, goes pair-bonding; some temporary, for the rearing of the young, some continuing for longer. In some species, this second purpose extends into a third; the bonding of a community and the imposing of a pattern upon its life. In man, all three elements are present, developed in combination to a unique degree. All three have a place, perennially, in a sexual morality; the balance between them varies with knowledge, place and time.

FERTILITY IN THE OLD LORE AND THE NEW

The old lore saw man as the bearer of seed (*semen*) and woman as the field in which he sowed it. If no child were conceived, it must have been the woman's fault: the field was barren. Moralities that fostered fecundity – in societies needing children for population growth, or to perpetuate a family 'name', an inheritance – permitted the childless wife to be repudiated, and another wife to be taken. The new lore, the discovery of oocytes and ovulation, revealed the falseness of the agricultural model. The discovery of spermatozoa within the seminal fluid, and the recognition of oligospermic or azoospermic seminal fluid, opened the possibility that childlessness might be attributed to the man instead of the woman. The old morality, derived from the old lore, is then seen to be unjust. The discovery of the Y chromosome, and with it the knowledge that it is from the male donation that the sex of the offspring is determined, would similarly reverse a morality that permitted the repudiation of a wife who bore only female children in societies that favoured males. Similarly, as knowledge unfolds of the endocrine and hormonal influences on the emerging embryo, resulting sometimes in degrees of variance between chromosomal sex and the 'sex' imprinted on the brain or as shown in the genital and other secondary characteristics, new attitudes must emerge towards persons whose behaviour has hitherto been misunderstood and, probably, condemned. Moral judgements are no substitute for the determining of fact.

The perennial sexual morality gave primacy to offspring; necessarily so in a world of high mortality where the whole economy – hunting, felling, tilling, harvesting, herding, fishing, fighting, craft and trade – called for population growth. Even then the primacy was not absolute; it has been subject to other claims. Religion did not invent, still less impose, marriage and the family; they emerged as biological and social necessities. Some religions, including the Jewish and the Christian, have invested marriage and the family with theological significance, and so have provided religious sanctions for a natural, biologically and socially grounded, morality. That morality has taken full account of the procreative need and exalted it: in the literature of the Old Testament, the Jews express delight in the blessings of womb and breast; in the enjoyment of motherhood, of paternity, of children and life in families. But because it is a human delight that is celebrated, the procreational thrust of the morality is humanly qualified. The second and third biological purposes of sex – pair bonding and the bonding of a community – are brought in alongside it; that is, *humanized*. So, while the Jewish law continued to permit divorce and polygamy, the prophetic voice pronounced it as a second best; a severance of the personal unity, becoming 'one flesh', which marriage in its essence was understood to be. So Malachi could champion, against a repudiating husband 'the wife of thy youth'; 'she is thy companion, and the wife of thy covenant'; 'I hate putting away, saith the Lord'.

A balanced sexual morality, therefore, can neither suppress the reproductive function of sexuality, nor make it an absolute. Reproduction is a humanized biological function maintained within a human context, a human couple 'bonded' by their sexual union and a human society built on kinship, the social recognition of family relationships. This human, that is, moral, limitation is relevant both to the promotion of fertility and to the regulation of it.

It is true that for about a thousand years there was a strain in Western Christian morality which would have given absolute primacy to the procreative intent. It was a false lead – like those that have come into science or philosophy or politics from time to time – brought into the Christian tradition from a pagan, Manichaean, source, and established there by St Augustine. He was, as Peter Brown sees him, the towering, outstanding intellectual and spiritual force of his generation, and so his teaching on this matter gained undue credence from his general authority. He taught that the only justification for venereal pleasure, that which saved it from being a sinful indulgence, was the *bonum prolis*, the intention to procreate children. That was an aberration from a tradition inherited from the Jews, of the wholeness of sexual union within an integrated relationship, and an aberration that Christianity, since the sixteenth century, has formally and now universally repudiated. In the centre of the Old Testament stands the Song of Songs, a cycle of poems in the form of a duet of two lovers with chorus, celebrating erotic love in all its naked beauty. These songs were echoed down the ages, notably in the *Carmina Burana*, and stood as a witness against the Augustinian tradition and its hard manifestations in the moralists' manuals and penitential practice. In short, the morality which we encounter in the West sees man as a sexual person, his sexuality accepted as integral to his personality, and its genital expression having a qualified but not absolute procreative function.

PROMOTING FERTILITY

There are human, that is, moral, considerations governing the promotion of fertility as well as the regulation of it. The new lore gives us means of remedying infertility not known before. These new means require, and are creating, an appropriate new morality: the morality of a practice. Hormonal inducement of ovulation is one possibility; assisted insemination with a husband's semen, or artificial insemination from a donor, is another; fertilization *in vitro* is an established third; and more may develop. If the interest in fertility were absolute, the treatment of infertility by one of these means would be merely an obligatory application of a technique, not necessarily medical, once the basic science and means of its application had been mastered. In fact, practitioners like Robert Edwards and John Brudenell and their colleagues are clearly conscious of the ethical dimensions of their work, and are concerned to identify and articulate them. The ethics concern, not only the normal medical criteria of risk and benefit, but also respect for and the maintenance of the human relationship within which the child will be born – the consent of the parties, and an assessment of their personalities and of their commitment to the enterprise. There are ethical considerations, also, relating to the donor of semen and the conditions governing his selection.

The shaping of a morality to accompany these new medical possibilities for the promotion of fertility has not been tardy or conspicuously contentious, among serious thinkers and practitioners. It has developed more steadily in a moral tradition where practitioners and moralists work together, seeking to ground the morality in its empirical base, than in the magisterial traditions, notably the Orthodox Jewish and the Roman Catholic, where the method favoured, at least by the magisterial authority itself, is an *a priori* one; the application of traditional precepts and principles with only secondary attention to the moral claims emerging from the practice. Roman Catholic moral theology, as enunciated authoritatively from the Vatican, grants a narrowly qualified permission for assisted insemination with a husband's semen, but none at all for AID. Some Roman Catholic moralists however, being allowed a certain liberty of professional speculation, are more positive on both treatments, particularly the first, than their authorities. The purpose of the Jewish and Roman Catholic objection is valid enough: it is to safeguard the integrity of the personal sexual exchange as the locus for human begetting – a ground on which fertilization *in vitro* is almost certain to be authoritatively condemned once the pronouncement comes. The point of principle may be granted, but the personal union will remain the normal means of begetting for all except a very few people most of the time. The occasional act of medical intervention does not in itself invalidate or weaken the principle, particularly when in the ethics of the practice, concern for the mutual commitment of the couple is high. An ethics for the application of fertilization *in vitro* and embryo transfer to 'proxy motherhood' has yet to be worked on. In some respects it will resemble the ethics of AID; in others it will differ.

REGULATING FERTILITY

The new knowledge has opened new possibilities also for the regulation of fertility, for the separation of sexual union from its product in conception. Both the capacity and the will to do so are peculiarly human, according to the evidence adduced in earlier chapters of this book, despite the sex-play of other primates like chimpanzees. The human pair is free from the oestrous cycle which limits congress in other species – the limitations of the menstrual cycle are minimal in comparison with its capacities for positive exploitation. The human pair has developed a capacity and a liking for a frequency and prolongation of coition far beyond that of even their nearest kindred among the primates; and they have developed penile and clitoral structures which make this enjoyment possible. With the dawn of understanding, the means were found to separate coition from conception, when there was the will. One such occasion has left its stamp upon Western consciousness. Onan's sister-in-law, Tamar, was left a widow without child. It was Onan's duty to give her a child, to perpetuate his brother's inheritance and name (not his own). But in the event 'he spilled his seed upon the ground', so that she should not conceive (*Genesis* 38). And in the hands of St Augustine that act became the great offence, evil in the sight of God. Augustine erred, in transferring the offence from the motive to the means: it was the refusal of the Levirate custom that violated the early Hebrew morality; the means was incidental to the rebellious end. But throughout the centuries 'the sin of Onan' has been alleged against coitus interruptus, the most common means of birth control. And such was the authority of Augustine, it is this tradition that the Roman Catholic authorities have in mind when they claim that the wilful exclusion of the possibility of conception from marital union has always been contrary to nature and the moral tradition of the Church.

Modern science has offered better means. In principle, the means are either to ensure that no ova are about at the time of coition or to assure that they do not meet spermatozoa if they are, or to prevent the implantation of the ovum in the womb if fertilization has occurred.

A perennial morality

Within the first means there are subdivisions, each with its own moral significance. Knowledge of the fertile and infertile periods within the menstrual cycle and how to time these by calculation, or by thermal, chemical or tactile test (knowledge vital to the successful treatment of infertility) gives the human pair a liberty of choice whether or not to come together in the fertile period if they want a child. This is the one method of conception control allowed, and even enjoined, for Roman Catholics by the authorities of their Church: the papal pronouncement in *Humanae Vitae* 1968 (s. 11, 14), remains authoritative, despite the reasoned objections of responsible Catholic moralists both before its promulgation and after. As Delhaye and his colleagues explain, life is tolerable for faithful Roman Catholic couples who dissent in conscience, only because of the admitted diversity of interpretation put upon the encyclical by local ecclesiastical hierarchies and by pastorally concerned moralists and parish priests.

To moralists not subject to the Roman obedience, and to some who are, there is a certain lack of logic in this position. As early as 1930, Pope Pius XI hinted in *Casti Connubii* (s. 24) at a 'wider' end in marital union, the 'complete and intimate life-partnership and association', as well as a 'narrower' end, the procreation and education of children. In 1951 Pope Pius XII, in an address to Italian midwives, *Vegliare con Sollecitudine*, after repeating the authoritative condemnation of 'birth prevention' by unnatural means, declared the 'use' of the infertile period lawful, and gave the midwives the positive duty of teaching this use to the mothers in their charge. The Second Vatican Council, in 1965, in its Pastoral Constitution on The Church in the Modern World, *Gaudium et Spes*, developed the hint of *Casti Connubii* into a full treatment of the personal as well as the procreational aspects of marriage (s. 47ff), clearly and notoriously laying the foundation for a widening of the liberty of choice in this respect. The then Pope, Paul VI, explicitly withdrew the question from the Council's competence, and referred it to a commission of his own (s. 51). The recognition that sexual union between husband and wife has a unitive function, and not merely a procreative one, is firm in the highest authoritative Roman Catholic teaching. Yet the only means authoritatively allowed for choosing between these functions on any given occasion is this boon of science, the exploitation of the fertile period. And why? It can only be because of a doubt about its certainty: 'each and every marital act (*quilibet matrimonii usus*) must remain open to the transmission of life' (*Humanae Vitae*, s. 11). The illogicality was compounded when the encyclical called upon men of science to provide 'a sufficiently secure basis for a regulation of birth, founded on the observance of natural rhythms', success in which would surely take away beyond doubt that 'openness to the transmission of life' on which the encyclical insists.

In fact, the negative regulation of *Humanae Vitae* is to a large extent morally discredited, even among faithful Roman Catholics. The use of contraceptive means other than the infertile period is widely held to be only *malum quia prohibitum* (wrong because forbidden) not *malum in se* (inherently evil). The task of the Roman magisterium now is to frame a new encyclical expressive of the developing moral consensus among an articulate laity which cannot for long be ignored. The consensus is the stronger because it too affirms what *Casti Connubii* and *Humanae Vitae* alike strive to protect, the essentially personal, human nature of the sexual act: all the language employed about the 'unnaturalness' of 'artificial' means is protective in intent, to preserve from distortion both the act of union itself and the relationship of persons in which it is accomplished.

The contraceptive pill is, in principle, a means of extending the infertile period throughout the menstrual cycle, and so of ensuring the absence of an ovum to be fertilized over a chosen period. Granted the acceptance of a liberty in general deliberately to prevent conception, the choice of this means is determined by aesthetics, a simple preference for a non-barrier method, and by a balance of benefits and risks. As with embolism or general malaise, the risks vary for different persons

at different ages; they are for medical assessment. In the
long term the consequences of taking 'the pill', the most
widespread non-therapeutic medication of persons ever undertaken, are still unknown and unpredictable. Only the experience
of generations descended from mothers who have used the pill
will reveal whether this fundamental invasion of the biological
process has been achieved without serious ill effect. The world
has accepted it; in some regions population growth is already
under control because of its sheer convenience, and in areas of
rapid population growth it appears to be the simplest remedy for
a desperate demographic problem. The balancing of benefit
against risk is on the social scale as well as the personal.

STERILIZATION

The second means by which the absence of ova or of sperm can
be ensured at the time of coition is by sterilization of the woman
or man. This again is a product of new knowledge. Anciently
the method was castration. The eunuch had his place in society,
from the keeper of the harem to the *castrati* voices in the operatic
chorus or *capella* choir. But the price was heavy: castration
virtually destroys sexual potency as well as fertility and results
in other changes of personality as well. When Origen, a highly
esteemed theologian of the third century, voluntarily made a
eunuch of himself in literal obedience to the saying of Jesus in
Matthew 19: 12, he was condemned for so doing; freedom from
temptation was to be won, not be removing its source but by
overcoming it by grace. Rome, as recently as 1976, repeated the
strict prohibition of direct sterilization even for therapeutic or
preventive purposes, because 'it deprives foreseen and freely
willed sexual activity of an essential element'. Non-Roman
Christianity does not condemn it, but would rather work out the
ethics of its acceptance and of its practice. The Board for Social
Responsibility of the Church of England initiated serious
thought to the question in 1962. In secular morality sterilization
is widely accepted as a voluntary measure, despite its condem-

nation as a violation of human dignity when compulsorily imposed, as in the West by the Nazis on the Jews and others in Germany, and by some States on mentally defective persons in the USA, and in the East by the State of Maharashtra, with the backing of the central government, in India. In 1976, in an English court, Mrs Justice Heilbron granted an injunction to restrain a surgeon from sterilizing a pre-pubertal mentally handicapped girl on the grounds that the deprivation of a capacity for parenthood without consent was a gross violation of a basic human right, and that a less drastic and better remedy, adequate social care, was available. The ethics of the practice hinge on its voluntariness, the informed consent of the patient or of both partners to a marriage, the degree of risk of their being left childless by the loss of their existing children, the interest of a partner to a subsequent marriage if any, and the degree of responsibility attaching to the choice of what is in effect a practically irrevocable act of contraception. If medical science should succeed in its development of long-term but reversible sterilization, the ethics would call for reconsideration accordingly.

OTHER MEANS OF CONTRACEPTION

Contraception in its modern form became established when the only known means were the interposing of a barrier between ovum and sperm – made widely possible by the development of latex rubber – or the killing of sperm by spermicidal preparations. Opposition from the established medical profession, as well as from religion, was partly moralistic, from the association of the condom in particular with prostitution and other extramarital intercourse, and partly aesthetic. The opposition has now been largely overcome, in secular morality and in most religious moralities. The International Planned Parenthood Federation can now report family planning programmes sponsored by governments throughout the world, including Islamic, Hindu and Buddhist cultures, though what progress they will make in the newly orthodox Islamic States of Pakistan and Iran is unpredictable. In virtually all Christian traditions, including the Roman Catholic, a positive duty of responsibility is taught for deciding the number and frequency of the births; partly because of the threat of regional over-population in the world; partly because of the intrinsic value of children, each with a claim to adequate care and resources; and partly because of the new role envisaged for women, wider than (though embracing) that of physical motherhood. Roman Catholic teaching differs in that it denies the liberty of choice of means which other Christian teaching enjoins. Once more, modern understanding of the fundamental biology removes a scruple. While it was believed that 'life' was contained in the father's 'seed' there could be an appreciable scruple against 'wasting life' in the seed. An understanding that 'life' begins with the conjunction of two elements, ovum and sperm, and that there is a regular wastage of both in nature without human volition, deprives the scruple of its hold. Contraception thus becomes a moral possibility; the way is then opened to work out an ethics for its use.

There remains one area of moral ambiguity in modern contraceptive practice, in which the ethics are not yet fully explored. This concerns the use of the intrauterine contraceptive device (IUCD), the coil or the loop, and of hormonal compounds that take effect after coition and possible fertilization rather than before it. The effect of these forms of intervention is apparently to prevent the implantation of the blastocyst in the womb, and so to assure its discharge. In strict theory, if 'life' is taken to begin at fertilization, this process is an early abortion, the deliberate destruction of a human life already conceived. The secular morality is untroubled by this scruple: IUCDs are in widespread use. Moral philosophers and theologians who concern themselves with the empirical base, the facts of the case, and with the moral claims arising from those facts, are aware of a tension in their judgments still unresolved. On the one hand they have a will to maximize the protection accorded to human life at all its stages. On the other they are aware of the high rate of loss of fertilized ova without intervention – upwards of 30%

according to some estimates – and so they are driven to ask whether implantation rather than fertilization is the stage at which the physiological assurance of continuity is strong enough to claim moral support. Certainly few moralists of this sort are disturbed in conscience by the early abortifacient action of the IUCD (if that is what it is) as they are by abortion, not required by medical necessity, later in fetal growth. Without this sense of 'moral claim' they are reluctant to designate an activity as an important moral issue.

ABORTION

Not by accident but rather by design abortion has been excluded from a consideration of the morality of contraception. It is here admitted, in writing of the IUCD, that there cannot be, in strict logic, an *absolute* distinction between contraception and abortion, as though the one were the prevention of a life coming into being and the other the destruction of a life already in being. If implantation is chosen as the stage when the duty to protect the fetus is held to begin, there is an arbitrary element in the choice, though one with good physiological warrant: implantation gives the blastocyst a higher degree of certainty for a viable future than it had before, In fact, in the Western tradition of law and morality, the protection accorded to the fetus has increased with the development of the fetus itself, concluding logically in English law with the creation of a new offence, child destruction, at the point of viability. The same logic operates still, with the increase in medical means to enable premature babies to survive, in the pressure to reduce the legal point of viability from twenty-eight weeks to something like twenty. A government Advisory Group chaired by Sir John Peel, were firm in this recommendation.

Implicit in this tradition is the admission that the protection accorded to the fetus has never been, and cannot be, absolute. The fetus, as a nascent human being, is presumed to have a right to live. That presumption is rebuttable only if the fetus is itself an aggressor on the same right in another person, the mother: innocent life must not be taken. *Innocent,* in this context, has no reference to moral blamelessness, freedom from guilt. It means, in the Latin language in which our jurisprudence was formed, *doing no injury or harm.* The harm threatened by the fetus has to be grave, to justify its destruction, and one irremovable by no less drastic means. This is the logic that accepts the morality of therapeutic abortion in cases, medically rarer now, of grave threat to a vital interest of the mother, but rejects abortion as an alternative to contraception or as a resort when contraception fails. It insists that a less drastic remedy be found than the taking of 'innocent' life. It seeks to maximize the sacredness, the inviolability, of human life at every stage – an aim consistent with the growing moral revulsion from the ruthless destruction of human life all too prevalent in this time of political terrorism, revolution and counter-revolution and the brutalities of military and police dictatorships. Abortion violates the dignity of women, and is all too often an instrument of male exploitation. Secular moralities throughout the Western world, and in some Eastern cultures, notably the Japanese, accept abortion as a method of fertility control, and even in countries, like Great Britain, where the intention of the law is to limit the permission to cases of serious threat to maternal well-being or to cases of severe congenital handicap. But the moralist must record the degree to which this acceptance is at variance with the moral tradition, and in particular (though not exclusively) with the Christian belief in the dignity of human life enshrined in that tradition.

SEXUALITY AND HUMAN BONDING

The basic biological functions of sexuality, it will be recalled from earlier chapters, are reproduction, pair bonding, and the shaping and bonding of groups, communities, and societies. We pass now, therefore, from a consideration of the moralities attending reproduction to those of bonding: to the humanizing of a basic biological relationship.

The critical factor in this transition is in the development of the cerebral cortex and the vast new possibilities that it opened up. The cortex in man 'has taken over much of the control of sexual behaviour' (p. 131) from the hormonal action that predominates in non-primates and, generally, in other primates also. There are disadvantages to this, in that the environmental influences, including the pressures of modern society, to which the cortex is open, may have adverse consequences in sexual behaviour. Maybe this is related causally to some of the variants of obsessive intensity described by Richard Green in Chapter 3. It would follow that there is a social and a personal interest in the shaping and control of these environmental influences, in the sort of pressures that society can exert. That interest requires an ethics apt to favour the more beneficial possibilities of cortical influence on sexual behaviour and to minimize the adverse.

The cortex as developed in humans is, moreover, instrumental in this process. For in addition to those functions of receiving and of linking heard and seen information, the sensations of touch and smell and the like, there are parts of the cortex that enable interpretation of what is received, and so also judgment, volition and choice. Ethics is, in a sense, the science of choice. An ethics of sexual behaviour, therefore, concerns those choices open to man, by which he can order his given biological potential to his best advantage in personal and social relationships.

The evolved human achievement has been to subordinate sex to what we may now call human capacities. Sex has developed from a biological encounter, basically genital and procreative, into a means of personal, human relationship. To hold that advance, and to take it further, is the object of an ethics of sex. Such an ethics will not undervalue the element of physical sensation which is part of the basic biological 'given', but it will insist that that sensation is itself richer and more enriching when it finds its place in a fully human relationship: it has a relational end, and is not an end in itself. It is a powerful ingredient in erotic attachment. In the words of Basil and Rachel Moss: 'the yearning for the beloved, joyful sexual passion shared, delight in the partner through the body in all kinds of ways – touching, exchanging of glances, etc. – but especially in the full mutual exchange of genital sex . . . We are bound to question the reduction of the word "erotic" to mean simply that which effects sexual arousal of the body. Rather, sex without *eros* is sex depersonalized; sex infused with *eros*, *eros* fulfilling one's sexuality, goes with new and disturbing and creative possibilities of commitment to another.' An erotic attachment of this sort pervades and engages two whole persons: affections, intellect, interests, outlook, judgment, will. It invites and grows upon commitment and fidelity in commitment – for which the simplest word is love; and out of this context genital sex, genital sensation, is unable to do its bonding work; it is incomplete.

MUTUALITY AND CONSENT

The full sexual relationship between human beings makes, therefore, certain demands. In addition to commitment there must be mutuality based on consent. The history of marriage could, in a limited sense, be written in terms of progression towards mutuality and consent. Polygamy served (and may still serve, in isolated localities) a useful social purpose when women perennially out-numbered men, had no education, no means of independent economic support, no employment, no pension, no protection or means of existence except by attachment to a man. But it must, and does, give way to monogamy when those conditions no longer exist – when medical intervention increases differentially the survival rate of boys, when girls are educated as well as boys, when social and economic change offers women employment, economic independence and some freedom of choice. Monogamy makes mutuality, and equal partnership, possible as polygamy does not. Educated women want it, perceive it as a right: and educated men learn to prefer it so. Roger Short's emphasis in Chapter 1 that reinforced bonding between the sexes provides a more secure social environment for rearing the young, and so maximizes the transmission of acquired characteristics in the long period of dependency, with enormous evolutionary advantage to human kind, is of high ethical significance. The traditional morality has always insisted on the essential inter-connection of the unitive and the procreative functions of human sexuality, however differently from time to time it has struck the balance between them.

Workers in the planned parenthood movements throughout the world know how much exploitation, with attendant physical harm and personal misery, accompanies child marriage and too-early child-bearing. It is a major point of strategy in the joint campaigns against over-population and morbidity to raise the age of marriage and to deliver girls from those traditions of social and physical degradation, including the use of chastity belts and the practices of infibulation and genital mutilation, which deny them equal participation, true personhood, in their sexual relations with men. How culturally insular, in a world perspective, is the propaganda to reverse the trend in Britain, and, by lowering the age of consent again, to expose children, especially girls, to even more sexual exploitation!

There is always a tension in sexual ethics between perceived goals and environmental pressures, social, demographic, economic, against their attainment. Once mutual fulfilment, based upon a partnership in free consent, becomes possible, and is seen to contribute to human flourishing, its support and protection become an ethical norm or obligation for that society. The supports may be cultural and social; they may in time become legal. In some societies the law has been in advance of practice, and a long period of development ensues before practice comes into line. Mutuality has long been entrenched in the Christian understanding of marriage and sexual relationship. It was clearly stated by St Paul, that great and much misunderstood first theologian and moralist of the Christian tradition. In Chapter 7 of his First Epistle to the Corinthians, sent in answer, point by point, to questions raised by them, he wrote:

Let the husband render unto the wife her due: and likewise also the wife unto her husband. The wife hath not power over her own body, but the husband: and likewise also the husband hath not power over his own body, but the wife. Defraud ye not one the other, except it be by consent for a season, that ye may give yourselves unto prayer, and may be together again, that Satan tempt you not because of your incontinency.

St Paul's idiom may not be our own; but what he prescribed, he prescribed for man and woman equally; similarly with consent. The canon law of the Church, which shaped marriage throughout the West after the withdrawal of the rule of the Roman Empire, required free consent by both parties as a condition of valid marriage and continued to do so for centuries, while the powerful, for political, dynastic or financial reasons, bestowed their children in marriage like pawns. 'Where both partners do not consent there is no marriage. Therefore those who give maidens in marriage when they are in their babyhood do nothing, unless both children give their consent when they reach the age of discretion': so declared the bishops of the Province of Canterbury in 1175, in the presence of King Henry II who had just done what the canon law forbade. It is only by continuous pressure, tenacity to a principle, that liberties now taken for granted have become established.

OTHER SEXUAL LANGUAGES

Genital union is one form of expression or language of sexual relationship. It is not the only language; a rich human culture will evolve many languages for the affirmation of sex, of a man affirming his manhood towards a woman or vice versa. Formal courtesies, dancing, and the attachment, where appropriate, of a role to gender are examples. So are varieties of dress, dress for different occasions, different modes of relationship, e.g. one for work, one for a formal evening out, one for a casual evening at home, one for tennis or for the beach, one or none for bed. There is an inevitable tension in this language, a tension that holds up to the climax of physical union – when genitally canalized: the orgasm is the final breakdown of tension. Without the sustained tension there would be far less in the world's store of great music, drama, poetry, legend, literature. Early sex education, the teaching of an ethics of sex, which concentrates upon genital contact – with its attendant emphasis on contraception and/or abortion – is a narrowing form of indoctrination, a closing of possibilities which a rich human culture would keep open. In a time of heavy pressure towards sexual conformity, towards the cultivation of genital excitement *per se*, whether within marriage or without it, celibacy itself, the ultimate expression of sexual distance, of tension unrelieved in orgasm, is a valuable witness for human liberty: the liberty not to conform, not to marry, not to give oneself or be taken physically at all. The renunciation may be negative but it need not be. There are many celibate lives, of men and of women, utterly fulfilled by means of those other languages of sexual exchange, other than the genital, in true man–woman mutuality.

CELIBACY

Celibacy has not always been embraced with this in mind. St Paul, while commending marriage prudentially for those disposed towards it ('it is better to marry than to burn'), stated a very carefully guarded preference for remaining unmarried 'because of the shortness of the time'. He and his contemporary Christians believed that the Lord would come again soon, in their own lifetime, and that nothing mattered more than that they and the world should be ready for his coming. Marriage was a distraction from that task. For those who were married, however, and those who should yet marry, the union should be full and firm (I *Corinthians* 7). For the same reason today, the Roman Catholic Church requires celibacy of its priests and offers powerfully the vocation of voluntary celibacy in communities of monks and nuns. Other Christian communities offer the same vocations, but without compulsory celibacy for priests. In the centuries between, however, the Manichaean denial brought in by St Augustine and others did exalt celibacy as a repudiation of sexuality; marriage was a second best, for those who could not attain the highest. Since the time of the Reformation that distortion has been ever more widely repudiated.

VARIANT FORMS OF SEXUAL BEHAVIOUR

What has been sketched here is a human norm: patterns and standards of human relationship, validated widely in ordinary experience, attainable and enjoyed by men and women in varying degrees, and approbated by the conventions, and in some respects by the laws, of societies with a developed awareness of human liberty. The variant forms of sexual behaviour described by Richard Green in Chapter 3 are now matters more of physiological and psychological than of moral interest. The relevant moral principles are those of acceptance of the person, sympathy and tolerance, within a concern for the common good, harmful aggression against which however has to be restrained. When restraint comes into question, the balance of claims between personal liberty and the common good becomes a fine one.

Of all the variant forms, homosexuality has secured for itself predominant interest over the last 30 years. In so far as it is a moral issue, moral judgment is divided. A distinction between a homosexual bias in personality and homosexual genital activity is commonly accepted by informed people. It is common ground that the biased condition is morally neutral; the division occurs over homosexual acts. In the secular moralities, professedly liberal and legally tolerant of private activity between consenting adults, there is still enough latent hostility to homosexual relationships for the press and news media to exploit for sensational purposes or for the discredit of public figures. Given a salacious scandal the British public shows itself less liberal than it professes to be, a fact the more dangerous in that forgiveness is not a cardinal principle in the secular morality. Among the religious moralities there is division. In the newly orthodox Islamic states, homosexual acts are punishable by death, as is heterosexual adultery. In Christian morality – the history of which has been ably scanned by Sherwin Bailey and the current spectrum by Kosnik and his colleagues – there are those who accept *in toto* the secular liberal view, countenancing homosexual acts as genuine expressions of love in a committed relationship and using the language even of 'homosexual marriage'; and there are those who regard such acts as intrinsically disordered, culpability for which may vary from grave to little or none according to person, occasion, motive and circumstance – and the offence, like all others, being always forgivable. The last is the formal position of the Roman Catholic Church which, because of its tenacious fidelity to the theological fullness of the Christian Gospel of forgiveness, need not be over-drawn by compassion into condonation of what it continues to designate as sin, but retains ampler means of dealing with it. (The author of this chapter is not a Roman Catholic.) Certainly, in terms of the basic biology, homosexual congress cannot fulfil the primary purpose of reproduction; in so far as it has pair bonding value and is pursued as such it is still, for the peace of all, dependent upon the tolerant goodwill of society.

Granted man's understanding of himself and of his place in the evolutionary scale, granted the biological 'given' together with that precious yet ambivalent possession, the developed cerebral cortex, with all its capacities both for distortion by environmental stimuli and for the expression of transcendent human powers, granted the variety of demographic, climatic, economic and social circumstance in which man lives, there can be no one universal pattern of human sexual behaviour. There must be variety and acceptance of variety. Yet there must be norms withal, conditions that make for human fulfilment in the widest possible way, to which men and women continually aspire. It is the function of ethics to discern what these are, and to foster conditions in which they may be attained.

A perennial morality

SUGGESTED FURTHER READING

Homosexuality and the Western Christian Tradition. D. S. Bailey. London; Longman Green (1955).

The Man–Woman Relation in Christian Thought. D. S. Bailey. London; Longman Green (1959).

Sterilization : an Ethical Enquiry. Board for Social Responsibility of the Church of England. London: Church Information Office (1962).

Augustine of Hippo : a Biography. P. Brown. London; Faber (1967).

Artificial Insemination. J. M. Brudenell, A. McLaren, R. V. Short and E. M. Symonds. London; Royal College of Obstetricians and Gynaecologists (1977).

Pour relire 'Humanae Vitae'. Declarations episcopales du monde entier. P. Delhaye, J. Grootaers and G. Thils. Grembloux; Duculot (1970).

The Artifice of Ethics. G. R. Dunstan. London; SCM Press (1978).

Fertilization of Human Eggs *in vitro.* Morals, Ethics and Law. R. G. Edwards. *Quarterly Review of Biology* **49,** 3 (1974).

Human Sexuality : New Directions in Catholic Thought. A. Kosnik, W. Carroll, A. Cunningham, R. Modras and J. Schulte. London; Search Press (1977).

Humanity and Sexuality, p. 16. B. Moss and R. Moss. London; Church Information Office (1978).

The Morality of Abortion : Legal and Historical Perspectives. Ed. J. T. Noonan. Cambridge, Mass.; Harvard University Press and London; Oxford University Press (1970).

The Use of Fetuses and Fetal Material for Research. Sir John Peel. Report of an Inter-Departmental Working Party. London; Her Majesty's Stationery Office (1972). (Attempted amendments to the Infant Life (Preservation) Act 1929 and of the Abortion Act 1967 are based on this Report.)

Index

Index

exhibitionism, 22, 23, 86, 89

fertility, in the old lore and the new, 153–5
  promoting (morality of), 155, 156
  regulating (morality of), 157
fertilization *in vitro* (human), morality of, 155, 156
follicle stimulating hormone, 71

gender identity, possible hormonal influence on, 62
genetic determination, of homosexuality, 61, 71
  of levels of sexual activity, 51
genitalia, female, development of the, 24–9
gibbon, 29, 128, 136
gonads, size of, 2–11
gorilla, 1, 3, 5–9, 11, 12, 14, 24, 28, 29, 31, 125, 126, 132, 149
group marriage, 95, 96
gull, herring, 136

hamster, 136
homosexuality, 1, 59–62, 64, 65, 68–75, 78, 79, 81–4, 86, 90, 91, 93, 133, 148, 149, 170, 171
  as abnormal sexual differentiation, 61–3
  hormones, role of, in human sexuality, 51–8

inbreeding, effects of, 138–42
incest, and paedophilia, 90–3
  guards against, 134, 136–48
  practice of, 136–8
indecent exposure, as male offence, 22, 23
indri, 128
infertile period, morality of use of, for birth control, 158–60
insemination, artificial, morality of, 155, 156
intercourse, adolescent, 100–2
  extramarital, 130, *see also* 'swinging' and group marriage
  premarital, 110–13, 130
intrauterine contraceptive device (IUCD), morality of use of, 162, 163

jackdaw, 136

kangaroo, 132
kibbutz, influence of social conditions in, 96, 144, 145
Kinsey seven-point scale of sexual partner preference, 69
kittiwake, 128

lion, 136
luteinizing hormone, 71
luteinizing hormone releasing hormone (LHRH), 72

Macacus arctoides, 149
macaque, 29
  Japanese, 146, 149
male sexual response, refractory period in, 46, 47
mammary cancer, 27, 28
marmoset, 29
marriage, 125–30
  group, 95, 96
  relatedness in partners, 136–48
masturbation, 63, 83, 86, 92
mating, assortive, 134, 135
maturity, sexual, preference for, for reproduction, 133–4
measurement of genital responses in laboratory, 35, 37, 40, 41
menstrual cycle, sexual activity during, 54–7
monkey, pigtail, 149
  rhesus, 54, 135, 146, 149
  spider, 149
  vervet, 149
monogamy, 1, 29–31, 32, 128
  serial, 31
morality, of abortion, 163, 164
  of artificial insemination, 155, 156
  of celibacy, 169, 170
  of embryo transfer, 156
  of fertilization *in vitro*, 155, 156
  of promoting fertility, 155, 156
  of regulating fertility, 157–64
  of variant forms of sexual behaviour, 170, 171
  sexual, 151–72
mouse, deer, 146, 147
mutuality and consent, 166–8

Index